Third Examination (B)	157
Answer Set	158
Third Examination (C)	161
Answer Set	162
Third Examination (D)	164
Answer Set	165
Third Examination (E)	167
Answer Set	170
Final Examination (A)	171
Answer Set	173
Final Examination (B)	177
Answer Set	179
Final Examination (C)	184
Answer Set	186
Final Examination (D)	190
Answer Set	192
Final Examination (E)	195
Answer Set	200

Organic Chemistry
Problems and Solutions

SECOND EDITION

Organic Chemistry
Problems and Solutions

Arthur D. Baker
Robert Engel

Queens College
The City University of New York

Allyn and Bacon, Inc.
Boston • London • Sydney • Toronto

Copyright © 1983, 1978 by Allyn and Bacon, Inc., 7 Wells Avenue, Newton
Massachusetts 02159. All rights reserved. No part of the material
protected by this copyright notice may be reproduced or utilized in any
form or by any means, electronic or mechanical, including photocopying,
recording, or by any information storage and retrieval system, without
written permission from the copyright owner.

Library of Congress Cataloging in Publication Data

Baker, Arthur D.
    Organic chemistry.

    1. Chemistry, Organic--Problems, exercises, etc.
I. Engel, Robert. II. Title.
QD257.B25 1983   547'.0076   83-2612
ISBN 0-205-07987-3

Printed in the United States of America.

10 9 8 7 6 5 4 3 2 1   88 87 86 85 84 83

CONTENTS

FIRST SEMESTER Page

First Examination (A)   1
   Answer Set   2

First Examination (B)   6
   Answer Set   8

First Examination (C)   14
   Answer Set   15

First Examination (D)   18
   Answer Set   19

First Examination (E)   21
   Answer Set   24

Second Examination (A)   25
   Answer Set   27

Second Examination (B)   30
   Answer Set   31

Second Examination (C)   35
   Answer Set   37

Second Examination (D)   40
   Answer Set   42

Second Examination (E)   43
   Answer Set   46

Third Examination (A)   47
   Answer Set   49

Third Examination (B)   52
   Answer Set   54

Third Examination (C)   57
   Answer Set   59

Third Examination (D)   62
   Answer Set   64

Third Examination (E)   66
   Answer Set   69

Final Examination (A)	70
Answer Set	73
Final Examination (B)	78
Answer Set	81
Final Examination (C)	87
Answer Set	90
Final Examination (D)	96
Answer Set	100
Final Examination (E)	103
Answer Set	111

SECOND SEMESTER

First Examination (A)	112
Answer Set	113
First Examination (B)	117
Answer Set	118
First Examination (C)	121
Answer Set	122
First Examination (D)	125
Answer Set	126
First Examination (E)	128
Answer Set	131
Second Examination (A)	132
Answer Set	133
Second Examination (B)	136
Answer Set	137
Second Examination (C)	140
Answer Set	141
Second Examination (D)	145
Answer Set	146
Second Examination (E)	148
Answer Set	151
Third Examination (A)	152
Answer Set	153

INTRODUCTION

This book of problems is intended to be used in conjunction with a standard organic chemistry text book. It is not intended to supplant the text either in the presentation of fundamental course material or as a source of numerous basic practice problems. Rather, it is intended as a source of review material and of problems that require a concerted application of concepts from a number of different parts of the organic chemistry course. It is organized in an examination format; we feel that this provides the student with practice in working under "pressure" as is common not only in chemistry class work, but also in "real" work in the world of chemistry. Answers are provided for all problems, many of these answers being in-depth analyses of the salient points.

We should make some note here of the general nature of organic chemistry, the proper use of this book by the student, and the nature of examinations in organic chemistry. Unfortunately, or fortunately, depending on one's point of view, the beginning student of organic chemistry finds that the subject matter differs from many other disciplines in its cumulative nature. That is, there is an interdependence of all topics developed throughout the course; separation of the topics beyond the pedagogically convenient formality of discussing each functional group in its own chapter is not realistic for the practice of organic chemistry. As the srudent will discover, the chemistry of alcohols remains significant when the chemistry of amino acids is under consideration. Organic chemists are generally concerned with molecules bearing more than one functional group. Therefore, one is forced to worry about the comparative reactivities of several groups simultaneously in planning a synthesis or analyzing a reaction. The problems given herein reflect this interdependence, particularly those included in the "First Semester Final Examination" section and beyond.

It is our intent that the problems in this book be used to assist the learning process in several ways. First, they should serve as a review of the course material by providing exemplary exercises incorporating the fundamental theory and reactions of organic chemistry. Second, the variety of problem types with their detailed explanations should assist in developing approaches toward the solving of "real" chemical problems. Third, problems incorporating a variety of types of chemical and physical data illustrate the interdependence of topics. Finally, the presentation of problems in an examination format introduces students to the pressures of solving problems with limitations of time and resources. Such limitations may seem artificial and applicable only to examinations; however, in "real" situations. chemists who must refer to the literature constantly for answers to simple questions or require excessive time to think find themselves at a distinct disadvantage.

Addressing ourselves at this point directly to the students who use this problem book, we recommend that you turn to it only after you have worked the problems in your text book, having thereby gained a fundamental understanding of the material at hand. To make the best use of the organization we have used, you should observe the recommended time limits given. While detailed answers are provided at the end of each examination, you should not turn to these until you have spent a considerable period of time attempting to solve them. A "considerable period of time" means possibly three times the period recommended. If you encounter difficulties, you should first turn to the appropriate section of the text book and again attempt to solve the problem. Turning to the answers too quickly leads to a rather superficial knowledge, and you do not learn to handle details as is required in organic chemistry. Looking over the answers to earlier examination should of course provide a useful review of material when you are preparing for the later examinations.

You may find the problems included here to be different in style from those used on examinations at your college or university, or by your instructor. We have attempted to use a variety of problem types taken from our actual examinations over the past few years. However your instructor might prefer only one or two of these types. Additionally, you may not cover exactly the same topics as we have done in our courses at Queens College, or in the same depth. In spite of all these possible differences from school to school, you will still find it useful to try all the problems given here.

A "good" examination will, in general, have problems of varying difficulty; this allows discrimination of the abilities of the students. A "good" examination will also approach being comprehensive, although a truly comprehensive one would be excessively long. Generally you will find the material not covered on an hour-long examination during the semester is featured prominently on the final examination.

Used properly, this problem book should assist you in the learning and performance of organic chemistry, at least in the sense of thinking about it. It should be remembered that the ultimate performance of organic chemistry occurs in the laboratory. No matter what one writes, on an examination or in a research grant proposal, it is relatively meaningless unless it has, or leads to, a reality in the laboratory.

ORGANIZATION

This book is divided into two parts, each corresponding approximately to one semester's work in a conventional university organic chemistry course. For each semester's work, four sets of examinations are provided. The first set (designated "First Examinations") in each case covers approximately the first one-thirs of a semester's work, the second set ("Second Examinations") covers the middle third, and the third set ("Third Examinations") covers the final third. The final set of examinations is intended to cover a whole semester's work, and is consequently labelled "Final Examinations". Thus an examination in the first semester labelled "Third Examination (D)" would be the fourth in a set of five examinations based on the last third of the first semester's work.

TEXTBOOK INDEX

In an effort to make this problem-solving manual more useful to the organic chemistry student who might be using any one of a variety of textbooks, we are providing here an index of the sample examinations. Each set of problems presented in an examination format is indexed with regard to applicable chapters for five of the more widely used organic chemistry textbooks. Consultation of this index will allow the student to determine which problem sets most closely correlate with the course topic material when any of the five textbooks are being used.

The five textbooks are: (a) "Organic Chemistry, 4th Edition", Morrison and Boyd; (b) "Introduction to Organic Chemistry, 2nd Edition", Streitwieser and Heathcock; (c) "Organic Chemistry, 2nd Edition", Solomons; (d) "Organic Chemistry", Wingrove and Caret; (e) "Organic Chemistry, 2nd Edition", Fessenden and Fessenden.

First Semester

Textbook Chapters

Examination	(a)	(b)	(c)	(d)	(e)
First A	1-3,7	2,3,5,10,11 15,19	1,3,5,6,14	1-4,7	1,2,4-6,9
First B	1-4,7	2,5-7,10,11	1,3,4,6,8	2-4,6,7	1,3,4,6,9
First C	1-4	2,4-7,15,19	1-4,8,14	2-6	1-4,6
First D	1-3,7	2,3,5,6,10 11,15,19	1,3-6,14	1-4,7	1-3,5,6,9
First E	1-4,6,7	2,4-7,10,11 15,19	1-6,8,14	2-7	1-6,9
Second A	5-8,13	4,5,10-12	3,5-7,9	3,5,7,8,11	4,5,9
Second B	4,6-8,13	7,10-12	2,6-9	6-8,11	4,9
Second C	3,4,6-8,13	6-8,11,12	4,6-9,14	4,6-9,11	4-6,9
Second D	4,5,7-9	5,7,8,11,20	3,5,7,8,10	3,6-8	3-5,9
Second E	1,3-8,10,13 14	2,3,5,7,10 11,12,20,22	1-3,6-11,15	2,3,6-8 10-13,19	1,3-5,7,9,10
Third A	6-8,10-12,15	4,8-11,18,19 23	5-7,12,13,15	5,7,10,14,15 19	5,7-10
Third B	7,8,10,11 15-17	9-11,14,22 23	6,7,11-13,15	7,8,10,12,14 15	7-10

Textbook Chapters

Examination	(a)	(b)	(c)	(d)	(e)
Third C	6,7,10,13,15 17	8-12,23	5,6,9,12,13 15	9-11,14,15	5,8-10
Third D	6,13,15,17	8,9,12,14 23	5,9,12,13	9,11,12,14 15	5,8-10
Third E	6,7,10,11 15-17	8-11,17 21-23	2,5,6,10,12 13,15,STH	9-12,14,15 18	5,7-10,21
Final A	1-4,6,8 10-13,15,17	2-4,6,8-12 14,15,19,23	1-5,7-9 12-15	1-6,8-12,15 19	1,3-10
Final B	2-4,6-11,15 17	4-11,14,15 19,20,23	2-8,10,12-15	2-10,12,14 15	1,3-10
Final C	1-12,15,17	2-5,7-11 13-15,17,19 23	1-3,5-8 12-16,STH	1-3,6-10,12 14,15,18,19	1,2,4-11,21
Final D	1-11,13,14 15,17	2,4,5,7-12 20,22,23	1-3,5-13,15	2,3,5-11 13-15	1-5,7-10
Final E	1-8,10-15,17	2-5,7-12,15 19,22,23	1-3,5,7-13 15	1-3,5-11 13-15,19	1,3-10
Second Semester					
First A	6,8,10-12 18-20	8,10,11,13 15,18,19	5,7,14-17	8-10,18,19 21-23	5-7,9,11-13
First B	5,8,10-12 18-20	10,11,13,15 18-20	5,7,14-17	8-10,19-23	7,9,11-13
First C	7,8,12,17-20	9-11,13 17-19	6,7,13,15-17 STH	8,10,15,18 19,21-23	7-9,11-13,21
First D	8,10-13 17-19	9-13,15,18 19	5,7,9,13-17	8-11,15,19 21-23	7-9,11-13
First E	6,10-13,16 18-21	10,12,13,15 18,19,22	5,7,9,12 14-17	9-11,14,19 21-24	7,9-13
Second A	18-24,26,32	13,15,18-20 24,25,27,30	15-18,20	10,17,22-24 26,27	7,11-15,17

Textbook Chapters

Examination	(a)	(b)	(c)	(d)	(e)
Second B	20-25,32	13,19,23-25 27,30	12,15-18,20	14,17,23,24 26	7,10,13-15
Second C	10,15,16,19 20-23	10,13,18,19 22-25	12,15-18	10,14,21-24 26	9-15
Second D	18,20-24	13,14,19,24 25,30	15-18	17,21,23,24 26	7,11,13-15
Second E	18-24,26,32 36	13,15,18,20 24,25,27,30	10,15-18,20	17,20-24,26 27	7,11-15,17
Third A	18,22,24,26 28,35	13,24,27,28 30,32	15,16,18-20 STO	17,21,24,26 29	7,14-16,18
Third B	8,20,28,30 32,35	11,19,20,28 29,32	7,10,17-20 STP,STO	8,23,24 26-29	9,13,14,16,18 19
Third C	12,18,19,21 23,26,28,30 32	10,13,18,20 24,27-29	10,15-20 STP	19,21,23,24 26-29	7,9,11,12,14 15,17-19
Third D	18,26,28,32 35	13,20,27,28 32	10,16,19,20 STO	21,24,26,27 29	11,14,16-18
Third E	21,22,24,28 30,32,33,35	13,20,24,28 29,32,34	10,16,18-20 STO,STP,STQ	24,26-29	14-19
Final A	10-12,17,19 20,22,26,28 30,33,35	9,10,14,15 17-19,24 27-29,32,34	13-15,17-20 STH,STO,STP STQ	10,12,15,18 19,21-24 26-29	6-8,11-19
Final B	10,12,15 17-21,23,24 30,33,35	9,10,13,14 17-19,21 23-25,29,30 32,34	10,12-18 STH,STO,STP STQ	12,14,17-19 21-24,26-28	7-17,19,21
Final C	12,15,17,20 21,23,24,28 30,32,33,35	9,10,13,14 20,21,23,25 28-30,32,34	10,12,13,15 16-19,STO STP,STQ	12,14,15,17 19,23,24 26-29	7,8,10 13-19,21
Final D	10-12,16-19 21-23,26,28 33,35	9,10,13-15 17-19,22,24 27,28,32,34	12,13,15-20 STQ	10,12,14,15 18-24,26,27 29	7-12,14-18 21

Textbook Chapters

Examination	(a)	(b)	(c)	(d)	(e)
Final E	10-12,16-19 21-23,26,28 33,35	9,10,13-15 17-19,22,24 27,28,32,34	12,13,15-20 STQ	10,12,14,15 18-24,26,27 29	7-12,14-18 21

Organic Chemistry
Problems and Solutions

FIRST SEMESTER

First Examination (A)
One Hour

1. (25 pts, 16 min) State the number of σ and π bonds in the folowing molecules and indicate the hybridization schemes involved with the starred atoms. Also show the geometries of the molecules.

    HC≡C*-CH=N*CH₃   H₂C=C*=CH₂
         1A   1B

2. (20 pts, 12 min) On combustion, 2.69 mg of an organic compound, $C_xH_yO_z$, gives 5.20 mg CO_2 and 3.20 mg H_2O. The compound is known to have a molecular weight of about 45 g/mole. What is its molecular formula?

3. (15 pts, 8 min) Starting with 2-chloro-2-methylbutane, ethyl chloride, and any inorganic compounds and solvents thought necessary, devise a synthesis of 3,3-dimethylpentane.

4. (20 pts, 12 min) Give structures for the major organic products of the following reactions and briefly explain their formation.

    (a) <u>tert</u>-butyl chloride + Mg/ether ⟶ (4A)

        (4A) + methyl alcohol ⟶ (4B)

    (b) $(CH_3)_2CBrCH_2CH_3$ $\xrightarrow{\text{KOH, ethanol, heat}}$

    (c) $(CH_3)_2CHCH(OH)CH_3$ $\xrightarrow{H_2SO_4,\ \text{heat}}$

5. (20 pts, 12 min) Draw the Newman projections for (a) the <u>gauche</u> conformation of <u>n</u>-butane and (b) the most stable conformation of 2,2,5,5-tetramethylhexane. Consider conformations about the center carbon-carbon bonds only, i.e. about the C_3-C_2 bond in (a) and the C_4-C_3 bond in (b).

First Examination (A)   Answer Set

1.

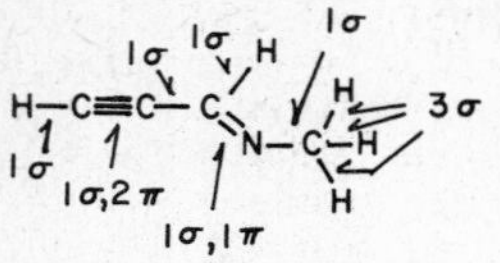

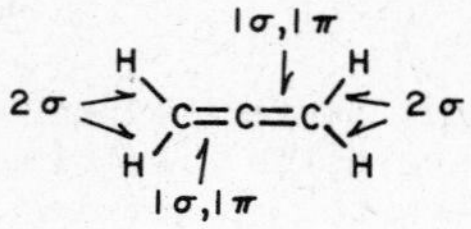

```
    1σ  1σ  1σ                          1σ,1π
     ↓   ↓H  ↓H                          ↓ H  ↓H
  H—C≡≡C—C*      3σ              2σ ⇀ C=C=C* ↽ 2σ
  1  1    1  N—C—H               H ↗  1    1   ↖H
 1σ 1σ,2π  1  ↑ ↑H                     1σ,1π
           1σ,1π
         1A                              1B
    Total: 9σ, 3π                    Total: 6σ, 2π
```

The important thing to remember in problems of this sort is that the <u>first</u>
bond between any two atoms is always σ; all other bonds between the same
atoms are π. There can <u>never</u> be more than one σ bond between any pair of
atoms.

For the second part of the problem one should recognize the general rule
that in organic chemistry π bonds are considered to result only from the
overlap of pure p orbitals, not from the overlap of hybrid orbitals, at
least when only first row elements are involved. Thus, considering molecule
1A, one sees that the starred carbon atom is involved in two π bonds, and
this must mean that two of its p orbitals are utilized in forming these π
bonds. This leaves only one p orbital to participate in the hybridization
process. The hybridization scheme is therefore <u>sp</u>, and the two hybrid <u>sp</u>
orbitals are used in forming the two C-C σ bonds to the starred carbon
atom. The starred nitrogen atom in 1A is involved in the formation of one
π bond, and so one of its p orbitals must be used in forming this bond.
This leaves two p orbitals to participate in the hybridization process used
for forming the σ bonds. The hybridization scheme is therefore <u>sp²</u>. Of
these three sp² hybrids, one accommodates the nitrogen "lone pair" and the
other two are used to form the two σ bonds involving the nitrogen atom.
The carbon atom in 1B is involved in the formation of two π bonds (one to
each of the terminal carbons in this molecule), and this requires that two
of its p orbitals be used in forming these π bonds. Only one p orbital
remains for the hybridization shceme which is therefore <u>sp</u>.

To decide on the geometries of the molecules, one must relate the
hybridization scheme to geometrical factors. The relationships one needs
are:

   <u>sp</u> hybridization results in linear geometry
   <u>sp²</u> hybridization results in trigonal geometry
   <u>sp³</u> hybridization results in tetrahedral geometry

To summarize the hybridization scheme for the various atoms in 1A and 1B:

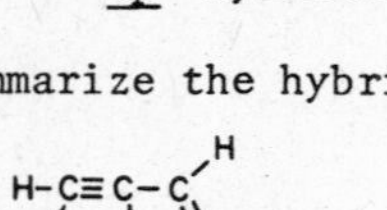

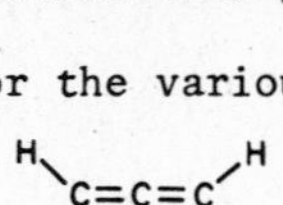

```
                  H                                  H       H
  H—C≡C—C                                             C=C=C
  1   1   1  N—CH3                              H        1     H
  sp  sp  1   1    3                                1  1   1
          sp²  sp                                  sp² sp  sp²
       1A                                              1B
```

Thus the geometry of 1A is:
```
                          120°  H
                     H—C≡C—C ↗ 120°
                           1   ↘ CH3
                         120° N
```

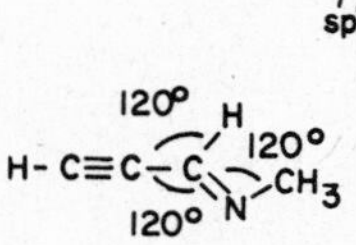

For 1B one must be careful to note that since the central carbon atom forms
π bonds to two different carbon atoms, the π bonds in this molecule must
be at right angles to one another because the p orbitals from which they
are formed are at right angles to one another. The net result then is two
doubly bonded systems perpendicular to one another.

2. All the carbon in the sample ends up as CO_2, and all the hydrogen as water.
   Thus: no. moles C atoms in sample = no. moles CO_2 produced, and
           no. moles H atoms in sample = 2 x no. moles H_2O produced.
   Remember that each mole of H_2O contains two moles of H atoms, and all
   hydrogen must originate in the sample.
   The first step, then, must be the calculation of the number of moles of C
   atoms and H atoms in the sample.

$$\text{no. moles C} = \frac{1 \text{ mole C}}{\text{mole } CO_2} \times 5.20 \text{ mg } CO_2 \times \frac{1 \text{ mole } CO_2}{44.0 \text{ g } CO_2} \times \frac{1 \text{ g}}{1000 \text{ mg}}$$

$$= 1.18 \times 10^{-4} \text{ mole C}$$

$$\text{no. moles H} = \frac{2 \text{ moles H}}{\text{mole } H_2O} \times 3.20 \text{ mg } H_2O \times \frac{1 \text{ mole } H_2O}{18.0 \text{ g } H_2O} \times \frac{1 \text{ g}}{1000 \text{ mg}}$$

$$= 3.55 \times 10^{-4} \text{ mole H}$$

One must now determine the weight of oxygen in the sample. This can be done
by comparing the total weight of C and H atoms with the given weight of the
sample (2.69 mg).

$$\text{wt. of C atoms} = 1.18 \times 10^{-4} \text{ mole C} \times \frac{12.0 \text{ g C}}{\text{mole C}}$$

$$= 1.42 \times 10^{-3} \text{ g C}$$

$$\text{wt. of H atoms} = 3.55 \times 10^{-4} \text{ mole H} \times \frac{1.0 \text{ g H}}{\text{mole H}}$$

$$= 3.55 \times 10^{-4} \text{ g H}$$

total wt. of C and H = 1.42×10^{-3} g + 3.55×10^{-4} g = 1.78×10^{-3} g

wt. of O in sample = 2.69×10^{-3} g − 1.78×10^{-3} g = 9.1×10^{-4} g

One must now compare the numbers of moles of C, H, and O.

no. moles C = 1.18×10^{-4}

no. moles H = 3.55×10^{-4} mole H

no. moles O = $\dfrac{9.1 \times 10^{-4} \text{ g O}}{16.0 \text{ g O/mole O}}$ = 5.7×10^{-5} mole O

The simplest ratio is found by dividing through by the smallest number (5.7×10^{-5}):

$$\text{C:H:O} = \dfrac{1.18 \times 10^{-4}}{5.7 \times 10^{-5}} : \dfrac{3.55 \times 10^{-4}}{5.7 \times 10^{-5}} : \dfrac{5.7 \times 10^{-5}}{5.7 \times 10^{-5}}$$

$$= 2 : 6 : 1$$

Thus the empirical formula is C_2H_6O. Since the molecular weight is only about 45, then the empirical formula must also be the molecular formula. The required answer is therefore C_2H_6O.

3. The conversion one must accomplish is

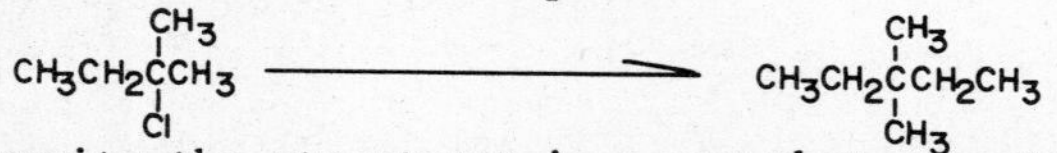

$$\underset{\underset{Cl}{|}}{\overset{\overset{CH_3}{|}}{CH_3CH_2CCH_3}} \longrightarrow \underset{\underset{CH_3}{|}}{\overset{\overset{CH_3}{|}}{CH_3CH_2CCH_2CH_3}}$$

First rewrite the structures in a way that accentuates a common structural feature in both compounds, a $CH_3CH_2C(CH_3)_2$ group. The required conversion is then

$$\underset{\underset{CH_3}{|}}{\overset{\overset{CH_3}{|}}{CH_3CH_2CCl}} \longrightarrow \underset{\underset{CH_3}{|}}{\overset{\overset{CH_3}{|}}{CH_3CH_2CCH_2CH_3}}$$

The problem therefore boils down to replacing a C-Cl bond with a $C-C_2H_5$ bond, and since ethyl chloride is one of the "allowed" reagents, this can be done by the Corey-House route shown below:

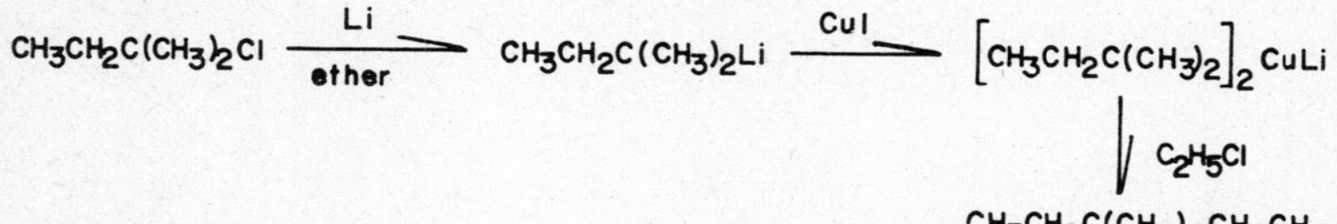

$$CH_3CH_2C(CH_3)_2Cl \xrightarrow[\text{ether}]{Li} CH_3CH_2C(CH_3)_2Li \xrightarrow{CuI} \left[CH_3CH_2C(CH_3)_2\right]_2 CuLi$$

$$\downarrow C_2H_5Cl$$

$$CH_3CH_2C(CH_3)_2CH_2CH_3$$

4. (a) The reaction of <u>tert</u>-butyl chloride with Mg/ether will give the corresponding Grignard reagent.

$$(CH_3)_3CCl + Mg \xrightarrow{\text{ether}} (CH_3)_3CMgCl$$

The reaction of Grignard reagents with almost any compound containing a hydrogen atom attached to oxygen results in the decomposition of the Grignard reagent and the formation of the corresponding alkane.

$$(CH_3)_3CMgCl \xrightarrow{CH_3OH} (CH_3)_3CH$$

(b) Heating an alkyl halide with KOH/ethanol results in dehydrohalogenation and formation of an alkene. An unsymmetric alkyl halide can result in a mixture of alkene products; however, the alkene containing the larger number of alkyl substituents generally will predominate.

$$\underset{\underset{Br}{|}}{\overset{\overset{CH_3}{|}}{CH_3\overset{}{C}CH_2CH_3}} \xrightarrow{KOH, \text{ ethanol}} \underset{\text{major}}{\overset{H_3C}{\underset{H_3C}{>}}C=C\overset{H}{\underset{CH_3}{<}}} + \underset{\text{minor}}{\overset{H}{\underset{H}{>}}C=C\overset{CH_3}{\underset{C_2H_5}{<}}}$$

(c) Treatment of an alcohol with hot sulfuric acid generally results in dehydration (loss of elements of water) to an alkene. The mechanism involves formation of an intermediate carbonium ion, which may rearrange. This is the case for the alcohol specified in this problem.

$$CH_3-\overset{\overset{H}{|}}{\underset{\underset{CH_3}{|}}{C}}-\overset{\overset{OH}{|}}{\underset{\underset{H}{|}}{C}}-CH_3 \xrightarrow{H^+} CH_3-\overset{\overset{H}{|}}{\underset{\underset{CH_3}{|}}{C}}-\overset{\overset{+OH_2}{|}}{\underset{\underset{H}{|}}{C}}-CH_3 \xrightarrow{-H_2O} \underset{2° \text{ carbonium ion}}{CH_3-\overset{\overset{H}{|}}{\underset{\underset{CH_3}{|}}{C}}-\overset{\overset{+}{|}}{\underset{\underset{H}{|}}{C}}-CH_3}$$

$$\downarrow \sim H^-$$

$$\underset{H_3C}{\overset{H_3C}{>}}C=C\overset{H}{\underset{CH_3}{<}} \xleftarrow{-H^+} \underset{\text{more stable } 3° \text{ carbonium ion}}{CH_3-\overset{\overset{+}{|}}{\underset{\underset{CH_3}{|}}{C}}-\overset{\overset{H}{|}}{\underset{\underset{H}{|}}{C}}-CH_3}$$

5. (a) The <u>gauche</u> conformation of <u>n</u>-butane may be written

[Newman projection showing CH3 groups adjacent]   or   [Newman projection showing CH3 groups adjacent]

or any other projection wherein methyl groups are adjacent.

(b) The most stable conformation of 2,2,5,5-tetramethylhexane (projection drawn about the C_3-C_4 bond) is

$$\underset{\underset{CH_3}{|}}{\overset{\overset{CH_3}{|}}{CH_3\overset{}{C}}}CH_2-\underset{\underset{CH_3}{|}}{\overset{\overset{CH_3}{|}}{CH_2\overset{}{C}CH_3}}$$
C_3—C_4 bond

[Newman projection with C(CH3)3 groups anti]

The bulky <u>tert</u>-butyl groups are placed as far away as possible from each other.

First Examination (B)   One Hour

1. (25 pts, 15 min) Consider the reaction of bromine monochloride, BrCl, with methane by a free radical chain mechanism. Two monohalogenated products would be possible according to the following overall equations:
    (i) $BrCl + CH_4 \longrightarrow CH_3Br + HCl$
    (ii) $BrCl + CH_4 \longrightarrow CH_3Cl + HBr$
   The following is a list of bond dissociation energies in kcal/mole:

CH_3-H	104
CH_3-Cl	84
CH_3-Br	70
$H-Cl$	103
$H-Br$	88
$Br-Cl$	52

   (a) Calculate ΔH for reactions (i) and (ii) and thereby deduce which would be most exothermic.
   (b) Write an equation for the chain initiating step for these reactions.
   (c) Calculate ΔH for the chain propagating steps for reactions (i) and (ii) without being concerned with processes leading to polyhalogenated species.
   (d) Predict whether the initial rate of formation of HCl will be faster or slower than that of HBr. Explain your answer.

2. (25 pts, 15 min) Draw an energy diagram for the various conformations of n-pentane (label axes "relative energy" and "degrees of rotation"). Consider rotation about the C_2-C_3 bond only. Indicate all maxima and minima for 360° of rotation about this bond and draw clear, concise Newman projections for the conformations considered.

3. (25 pts, 15 min) Answer all parts:
   (a) Account for the following reaction:

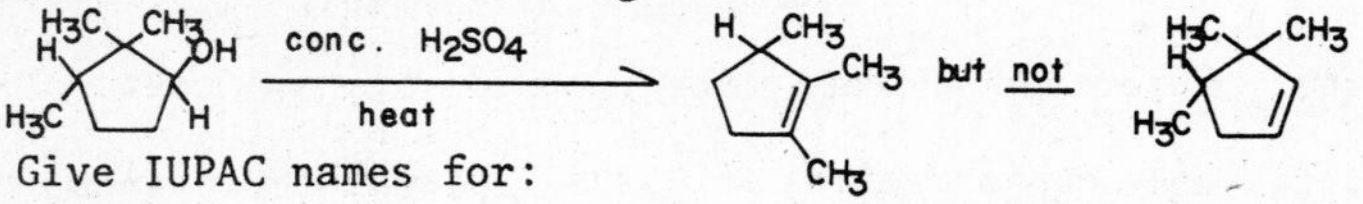

   [structure of 1,1,2-trimethyl-2-cyclopentanol with H3C, CH3, OH groups] $\xrightarrow[\text{heat}]{\text{conc. } H_2SO_4}$ [1,2,3-trimethylcyclopentene product] but not [alternative cyclopentene product]

   (b) Give IUPAC names for:

   $(CH_3)_2CHCH(CH_3)_2$

   $CH_3CH_2\underset{2}{\overset{\overset{\displaystyle CH_2CH_2CH_3}{|}}{C}}HCH(CH_3)CH_2CH_3$

   (c) Write structural formulas for (E)- and (Z)-1,2-dichloropropene and decide which will have the higher boiling point. Explain your answer.

4. (12 pts, 7 min) For each of the following write acceptable Lewis electron-dot representations, including significant resonance structures if appropriate.

   (a) cyanogen bromide   (b) the nitrite ion, NO_2^-

5. (13 pts, 8 min) Make the conversions between the Newman and Fischer ("cross") projections below and designate all chiral centers as either R or S.

First Examination (B)   Answer Set

1. (a) In reaction (i) a Br-Cl bond is broken (requiring 52 kcal/mole) and so is a CH_3-H bond (requiring 104 kcal/mole), while a CH_3-Br bond is formed (generating 70 kcal/mole) along with an H-Cl bond (generating an additional 103 kcal/mole). Thus a total of 156 kcal/mole is expended in breaking bonds and a total of 173 kcal/mole is generated by bond formation. The reaction is therefore exothermic by 17 kcal/mole.
    ΔH for reaction (i) = -17 kcal/mole
    (Note that exothermic reactions have a negative ΔH.)
    Proceeding similarly for reaction (ii),
    ΔH = (+52 + 104 - 84 - 88) kcal/mole
          = -16 kcal/mole

    (b) The chain initiating step for both reactions (i) and (ii) is the homolytic fission of the Br-Cl bond.
    $$Br-Cl \xrightarrow{h\nu \text{ or heat}} Br\cdot + Cl\cdot$$
    (c) Chain propagating steps for reaction (i) are:

    $Br\cdot + CH_4 \longrightarrow HBr + CH_3\cdot$ ΔH = (+104 -88) = 16 kcal/mole (1)
    and $Cl\cdot + CH_4 \longrightarrow HCl + CH_3\cdot$ ΔH = (+104 -103) = 1 kcal/mole (2)
    followed by:
    $CH_3\cdot + Br-Cl \longrightarrow CH_3Br + Cl\cdot$ ΔH = (+52 -70) = -18 kcal/mole (3)

    Chain propagating steps for reaction (ii) are:
    The first two reaction above, followed by

    $CH_3\cdot + Br-Cl \longrightarrow CH_3Cl + Br\cdot$ ΔH = (+52 -84) = -32 kcal/mole (4)

    (d) It is seen that the chain propagating steps (1) and (2) are endothermic and steps (3) and (4) are exothermic. Steps (1) and (2) are therefore more difficult, and hence rate determining, as long as they are true equilibria. Since (2) is much less endothermic than (1), it may safely be predicted that (2) will proceed more rapidly, so HCl will form faster than HBr.

2. n-Pentane may be written
    $$CH_3CH_2CH_2-CH_2CH_3$$
    which is equivalent to C_2-C_3 bond
    $$CH_3CH_2-CH_2C_2H_5$$

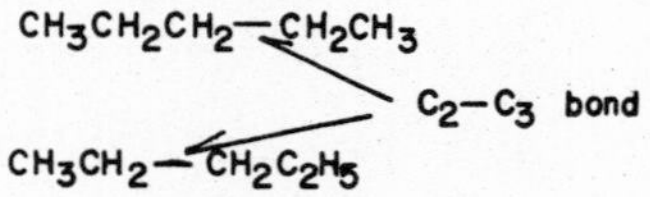

    Considering 360° of rotation about C_2-C_3, one may then construct the following energy diagram:

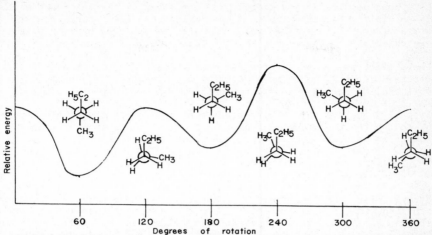

3. (a) One should recognize that there are various equivalent representations of certain structures, e.g.

Keeping this in mind the course of the reaction may be explained as follows

(b)

    $(CH_3)_2CHCH(CH_3)_2$ 2,3-dimethylbutane

        $CH_2CH_2CH_3$
        |
    $CH_3CH_2CHCH(CH_3)CH_2CH_3$ 4-ethyl-3-methylheptane

(The longest chain is 7 carbon atoms, so this compound must be named as a derivative of heptane.)

(c)

    (E)-   (Z)-

In the (E)- isomer, the bond dipole moments associated with the two C-Cl bonds will cancel and the molecule will therefore have only a small net dipole resulting from the slightly polar C-CH$_3$ bond. The (Z)- isomer on the other hand will have an appreciable net dipole moment because the two

C-Cl bond dipoles do not cancel; in fact, they partly reinforce each other. In general, a greater dipole moment implies greater intermolecular attractions and this is in turn indicated by a higher boiling point. Thus one expects the (Z)- isomer to have the higher boiling point.

4. Lewis electron-dot structures for BrCN and NO_2^- are required. Whenever such electron-dot structures must be drawn a systematic procedure should be followed. Described below is the procedure for the specified structures.
(a) BrCN
Step 1. Calculate the number of valence-shell electrons in the structure by adding the contribution from each atom, which is indicated by the number of its group in the periodic table.

Carbon	4
Nitrogen	5
Bromine	7
Total	16

Step 2. Place the total number of electrons given by step 1 around the atoms so that each has an octet. This is best approached in a number of stages. First, we know that there must be at least one pair of electrons between each pair of atoms, so a partial structure I may be drawn.

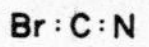

$$Br:C:N$$
$$I$$

There remain 12 electrons to place around the three atoms so that each gets an octet. In partial structure I bromine requires 6 more electrons, carbon 4, and nitrogen 6. Since this totals 16 electrons and there are only 12 available, a sharing of electrons is inevitable, implying the existence of multiple bonds in the final structure. In fact, either one triple bond or two double bonds appear necessary. Thus the structures II-IV are suggested.

:Br:::C:N: :Br::C::N: :Br:C:::N:
   II 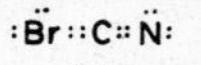   III 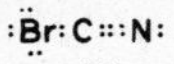   IV

Step 3. To decide which of these structures is the appropriate one formal charges must be assigned. For group A elements formal charges can be derived from the following equation:

FORMAL CHARGE = GROUP NUMBER -1/2 NUMBER OF SHARED ELECTRONS
                              -TOTAL NUMBER OF UNSHARED ELECTRONS

Thus for the bromine atom in structure I the formal charge is given by
    7 - 1/2(6) - 2 = 2
Following these rules the formal charges in structures II-IV are:

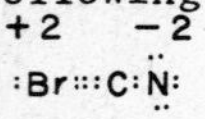

+2   -2   +1   -1
:Br:::C:N: :Br::C::N: :Br:C:::N:
   II 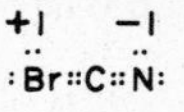   III 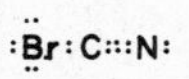   IV

Structure IV contains the least separation of charge and therefore is the most reasonable one. Thus the required answer is

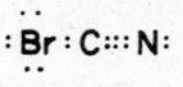

:Br:C:::N:

alternately represented by

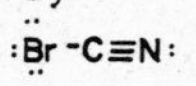

:Br̈-C≡N:

Sometimes the lone pairs are omitted and the structure is represented

$$Br-C\equiv N$$

This is done for the sake of simplicity; the lone pairs are still there. Their presence is understood.

(b) NO_2^-

<u>Step 1</u>. Proceeding as before, one finds the total number of valence electrons.

Nitrogen	5
2 Oxygen	12
Negative Charge	1
Total	18

<u>Step 2</u>. Write all reasonable structures with 18 valence electrons assuming that nitrogen is bonded to two oxygen atoms.

$$:\ddot{O}:\ddot{N}::\ddot{O}: \qquad :\ddot{O}::N:\ddot{O}:$$
$$\text{I} \qquad\qquad \text{II}$$

<u>Step 3</u>. Assign formal charges.

$$\underset{\text{I}}{:\overset{-}{\ddot{O}}:\ddot{N}::\ddot{O}:} \qquad \underset{\text{II}}{:\ddot{O}::N:\overset{-}{\ddot{O}}:}$$

Structures I and II are thus equivalent and the nitrite ion is best described as a hybrid of structures I and II. The required answer then is:

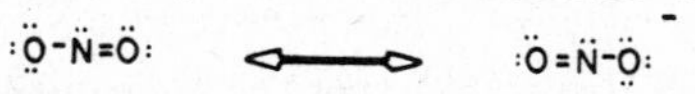

$$:\overset{-}{\ddot{O}}-\ddot{N}=\ddot{O}: \longleftrightarrow :\ddot{O}=\ddot{N}-\overset{-}{\ddot{O}}:$$

This can alternatively be represented as:

$$\overset{-}{O}-N=O \longleftrightarrow O=N-\overset{-}{O}$$

In such a case the presence of lone pairs is understood as in the previous example.

5. Fischer projections are, in effect, "flattened out" representations of eclipsed Newman projections. Thus to convert from a Newman to a Fischer projection these steps are followed: (1) rotate the groups to produce an eclipsed Newman projection; (2) "flatten out" this eclipsed Newman. For the particular example given, this works as follows:

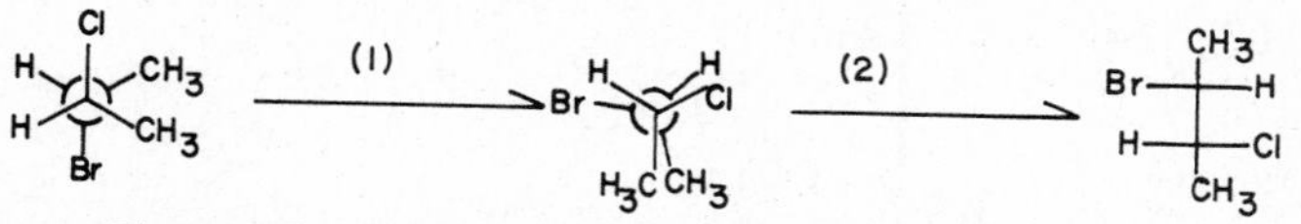

Note: The methyl groups are placed in the bottom position since the Fischer projection requires the methyl groups to be shown pointing away. Finally, <u>R</u> and <u>S</u> designations must be assigned.

```
         CH₃    R
    Br ──┬── H
         │  ╱   R
    H ──┼── Cl
         CH₃
```

Note: A special section on stereochemical representations follows this set.

To convert from a Fischer to a Newman projection the reverse procedure is followed. However, for the example given, there is the added complication that the acetylenic linkage is required on the <u>front</u> carbon atom in the Newman projection whereas it appears on the <u>back</u> carbon atom in the Fischer projection. To handle this difficulty one recalls that one can rotate a Fischer projection in the plane of the paper by 180° without affecting the stereochemical requirements.

```
          H                              CH(CH₃)₂
    Cl ──┬── C≡CH           H₃CCH=CH ──┬── H
    H  ──┼── CH=CHCH₃         HC≡C ────┼── Cl
       CH(CH₃)₂   identical to          H
```

Then, convert to an eclipsed Newman projection:

```
           H Cl                Finally, rotate groups to         C≡CH
            \│                 give required disposition:   H ──┬── CH(CH₃)₂
     HC≡C ──C── CH(CH₃)₂                                    H ──┼── Cl
    H₃CCH=CH  H                                                CH=CHCH₃
```

To decide on $\underline{R}$ and $\underline{S}$ designations for this problem one must use the priority rules. Using these rules one sees that

$$-C\equiv CH \text{ is equivalent to } -\underset{C}{\overset{C}{C}}-\underset{C}{\overset{C}{C}}-H$$

and

$$-CH=CHCH_3 \text{ is equivalent to } -\underset{C}{CH}-\underset{C}{CH}-CH_3$$

Thus for the two chiral carbon atoms the order in the substituent groups is

(1) $-Cl > -C\equiv CH > H\overset{|}{C}-CH=CHCH_3 > -H$
    $CH(CH_3)_2$

(2) $ClCHC\equiv CH > -CH=CHCH_3 > -CH(CH_3)_2 > -H$
    $|$

Thus the $\underline{R},\underline{S}$ designations are as follows:

```
                 H   R
                  \ ╱
         Cl ──┬── C≡CH
     S ─      │
         H  ──┼── CH=CHCH₃
            CH(CH₃)₂
```

SPECIAL NOTES ON STEREOCHEMICAL REPRESENTATIONS

It is helpful to remember that swapping any two groups attached to a chiral carbon necessarily alters the absolute configuration at that chiral center from R to S or vice versa. A second swap of two groups attached to the same carbon atom will return to the original configuration. This can be useful if it is necessary to rearrange groups in a stereochemical drawing to highlight any particular stereochemical feature. For example, if one wishes to decide on R or S assignments, it is necessary to have the group of lowest priority pointing away, and the stereochemical drawing as presented may not indicate this. Consider, for example, the following Fischer projection:

$$\begin{array}{c} Cl \quad \textcircled{1} \\ H—\!\!\!\!\!—Br \\ H—\!\!\!\!\!—Br \\ Cl \quad \textcircled{2} \end{array}$$

Consider first the carbon number 1. It is necessary to place the H atom in the top position if one wishes to make an R,S assignment. However, one can not simply make a swap of H and Cl since this would lead to a mirror-image arrangement of the groups about the carbon atom. Therefore one must make a double swap if one wishes to preserve the stereochemistry.

$$\begin{array}{c} Cl \\ H—\!\!\!\!\!—Br \\ H—\!\!\!\!\!—Br \\ Cl \end{array} \text{ is equivalent to } \begin{array}{c} H \quad R \\ Br—\!\!\!\!\!—Cl \\ H—\!\!\!\!\!—Br \\ Cl \end{array}$$

Thus one can see that carbon atom number 1 must be described as R absolute configuration. Similarly, carbon atom number 2 can be shown to have S absolute configuration.

$$\begin{array}{c} H \\ Br—\!\!\!\!\!—Cl \\ H—\!\!\!\!\!—Br \\ Cl \end{array} \text{ is equivalent to } \begin{array}{c} H \\ Br—\!\!\!\!\!—Cl \\ Br—\!\!\!\!\!—Cl \\ H \quad \underline{S} \end{array}$$

First Examination (C)   One Hour

1. (25 pts, 15 min) In the free radical reactions of alkanes with chlorine at room temperature, the relative reactivities of primary, secondary, and tertiary hydrogen atoms are in the ratio 1.0:3.8:5.0. Predict the relative proportions of all expected monochlorinated products from the reaction of 2,3,4-trimethylpentane with chlorine in the presence of light at room temperature.

2. (17 pts, 10 min) Draw a Fischer projection formula for 2R,3R-2-chloro-3-bromobutane, and also draw a Fischer projection formula for one of the diastereomers of this compound. For partial credit, clearly indicate the rank of the groups at each chiral center and indicate the R,S assignments in the diastereomer chosen.

3. (17 pts, 10 min) Outline methods for the preparation of
   (a) $(CH_3)_2CHD$ from propane
   (b) isohexane from propyl bromide and isopropyl bromide
   In addition to the specified starting materials, one may use any needed inorganic substances and any solvents.

4. (25 pts, 15 min)
   (a) The table following lists the melting points and boiling points of n-pentane and neopentane. Explain why the boiling point of n-pentane is higher than that of neopentane, while of the two compounds n-pentane has the lower melting point.

	mp	bp
$CH_3(CH_2)_3CH_3$	-130°	36°
$(CH_3)_4C$	-17°	9.5°

   (b) Calculate the total number of σ and π bonds, and note the hybridization scheme for the starred atoms in ketene, which has the Lewis structure

   $$\overset{H}{\underset{H}{>}}\overset{*}{C}=\overset{*}{C}=\overset{*}{\ddot{O}}:$$

   Also, deduce the geometry of the molecule.

5. (16 pts, 10 min) For each of the following equilibria is K_{eq} greater than 1 or less than 1? Offer a brief explanation in each case.
   (a) $HF + Br^- \rightleftharpoons HBr + F^-$
   (b) $CH_3OH + CH_3^- \rightleftharpoons CH_4 + CH_3O^-$
   (c) $CH_3SH + CH_3O^- \rightleftharpoons CH_3OH + CH_3S^-$
   (d) $CH_3MgCl + HCl \rightleftharpoons CH_4 + MgCl_2$

First Examination (C)

Answer Set

1. The first step is to pick out all chemically equivalent hydrogen atoms, i.e. hydrogen atoms which when replaced by chlorine atoms will give the same product. In this case there are four types of hydrogen atoms, and thus there will be four different monochlorinated products

```
                        Type d
                          ↓
         Type a → CH₃    CH₃     CH₃ ← Type a
                  \       |       /
           CH₃ — C ——— C ——— C — CH₃
                  |       |       |
                  H       H       H
                ↗         ↑        ↖
           Type b       Type c     Type b
```

Type of H replaced	Product	No. in starting cpd.
a, 1°	1-chloro-2,3,4-trimethylpentane	12
b, 3°	2-chloro-2,3,4-trimethylpentane	2
c, 3°	3-chloro-2,3,4-trimethylpentane	1
d, 1°	3-(chloromethyl)-2,3-dimethylpentane	3

The relative proprtions of the four products are now determined by multiplying the appropriate reactivity factors by the number of equivalent hydrogens.

For a   12 x 1.0 = 12.0
For b   2 x 5.0 = 10.0
For c   1 x 5.0 = 5.0
For d   3 x 1.0 = 3.0

Thus the ratio of the products will be a:b:c:d = 12:10:5:3.

2.
```
        H
    H₃C—|—Cl
    Br —|—CH₃
        H
```
Other equivalent Fischer projections can be drawn; see Special Notes on Stereochemical Representations.

The order of priority of the groups is

```
         ④
         H
    ③H₃C—|—Cl ①
       BrCHCH₃
         ②      for C-3
```

```
              ②
          CH₃CHCl
       ①Br—|—CH₃ ③
              H
              ④      for C-4
```

For the diastereoisomer either of the following is acceptable:

```
         H    S
     Cl—|—CH₃
     Br—|—CH₃
         H
         R
```

```
         H    R
     H₃C—|—Cl
     H₃C—|—Br
         H
         S
```

3. (a) CH₃CH₂CH₃ —Cl₂, hν→ CH₃CHCH₃ —Mg, Et₂O→ CH₃CHCH₃
                                  | |
                                  Cl   MgCl

followed by

15

$$\underset{\underset{MgCl}{|}}{CH_3CHCH_3} \xrightarrow{D_2O} \underset{\underset{D}{|}}{CH_3CHCH_3} + Mg(OD)Cl$$

(b) $\underset{\underset{Br}{|}}{CH_3CHCH_3} \xrightarrow{Li, Et_2O} \underset{\underset{Li}{|}}{CH_3CHCH_3} \xrightarrow{CuI} \left[(CH_3)_2CH\right]_2CuLi \xrightarrow{CH_3CH_2CH_2Br} (CH_3)_2CHCH_2CH_2CH_3$

The organometallic compound should be made from the isopropyl rather than the n-propyl bromide since the alkyl halide used in the final step should be primary.

4. (a) Melting points are heavily influenced by the symmetry of a molecule; more symmetric, more compact molecules fit better into a crystalline lattice making the lattice more stable. Neopentane, being more compact and symmetric than n-pentane, is therefore expected to form a more stable lattice than that of n-pentane. However, because n-pentane is a more "diffuse" molecule than neopentane, its electrons are more polarizable, giving rise to increased van der Waals attraction. The result is that n-pentane has the higher boiling point because of this increased amount of van der Waals attraction. In the crystalline state, even though this van der Waals force of attraction between neighboring n-pentane molecules exceeds that between neighboring neopentane molecules, the latter can exert an attractive force on more neighboring molecules than can the less symmetric n-pentane, which, as pointed out above, fits less well into a crystal lattice. The net result of many small neopentane/neopentane molecular interactions then exceeds the net sum of fewer but stronger n-pentane/n-pentane interactions, and this results in neopentane having the higher melting point.

(b) In each double bond there is one σ and one π bond. (Note: in a triple bond there is one σ bond plus two π bonds.) Thus the total number of σ bonds in the molecule is four (two from carbon to hydrogen, one from carbon to carbon, and one from carbon to oxygen).
Since there are two double bonds in the molecule, there must be two π bonds.
The first carbon atom forms one π bond, thus leaving only 2 p orbitals to participate in the hybridization scheme for the 3 σ bonds. The hybridization is therefore $\underline{sp^2}$, giving trigonal geometry around the first carbon.

$$\underset{H}{\overset{H}{}} \overset{sp^2}{\diagdown}C=C=O \qquad 120°$$

The second carbon atom forms two π bonds. As has been seen in earlier examples, this results in sp hybridization and linear geometry.

$$\underset{H}{\overset{H}{}}\diagdown C \overset{sp}{=} C = O$$

The oxygen atom forms one double bond. This means that it must use an $\underline{sp^2}$ hybridization scheme. Presumably the oxygen atom uses hybrid orbitals rather than unhybridized ones in order to form the strongest possible bond to carbon. Only one of the three sp² hybrid orbitals is needed to form a bond; the other two accommodate two lone pairs.

This sp² orbital forms a bond to the central carbon atom by overlapping with a C sp orbital.

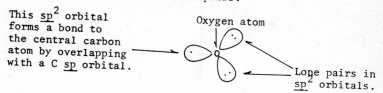

Oxygen atom

Lone pairs in sp² orbitals.

5. Parts (a)- (c) are acid-base reactions. Thus the equilibria will favor the weaker acids and bases.

(a)   HF   +   Br⁻   ⇌   HBr   +   F⁻
     Weaker   Weaker   Stronger   Stronger
     acid   base   acid   base
     K_{eq} less than 1

(b) CH_3OH + CH_3^- ⇌ CH_4 + CH_3O^-
     Stronger   Stronger   Weaker   Weaker
     acid   base   acid   base
     K_{eq} greater than 1

(c) CH_3SH + CH_3O^- ⇌ CH_3OH + CH_3S^-
     Stronger   Stronger   Weaker   Weaker
     acid   base   acid   base
     K_{eq} greater than 1

Part (d) is answered easily if one realizes that Grignard reagents are unstable in even mildly acidic media. Clearly then, K_{eq} for (d) is appreciably greater than unity.

First Examination (D)   One Hour

1. (25 pts, 15 min) (a) 2-Chlorobutane was formed in low yield in the reaction of ethane and chlorine at 400°. By means of equations show how it was formed. Show all steps clearly.
   (b) For the monochlorination of 3-methylpentane using chlorine and light, write the IUPAC names and draw the structures for all monochloro isomers produced and predict the relative percentages of the isomers, assuming the relative rates of abstraction of 3°:2°:1° hydrogen atoms are 5:3.8:1.

2. (25 pts, 15 min) (a) Write equations for (i) the homolytic, and (ii) the heterolytic cleavage of the S-S bond in hydrogen persulfide, H_2S_2. Show clearly whether the products are positively charged, negatively charged, free radical, or otherwise.
   (b) Given the following table

	ΔH, kcal/mole			
	X = F	Cl	Br	I
(1) $X_2 \longrightarrow 2 X\cdot$	38	58	46	36
(2) $X\cdot + CH_4 \longrightarrow HX + CH_3\cdot$	-32	1	16	33
(3) $CH_3\cdot + X_2 \longrightarrow CH_3X + X\cdot$	-70	-26	-24	-20

   Estimate the following. If there is insufficient data to do so, so state.
       (i) The value of E_{act} for $Br_2 \longrightarrow 2 Br\cdot$
       (ii) The value of E_{act} for $2 I\cdot \longrightarrow I_2$
       (iii) Which halogen is most reactive to CH_4
   (c) Show on an energy progress diagram the conversion of CH_4 to CH_3Cl by the reactions depicted in equations (2) and (3) above.

3. (25 pts, 15 min) Provide structures and IUPAC names for the following:
   (a) an alkane C_9H_{20} which can form only two different monochlorinated products.
   (b) a compound of molecular weight 114 which contains 15.8% hydrogen and 84.2% carbon and which gives a single compound when brominated under free radical conditions.
   (c) the major product obtained from the reaction:

   $$(CH_3)_3CCH(OH)CH_3 \xrightarrow{\text{conc. } H_2SO_4}$$

   (d) the product resulting from the following sequence:

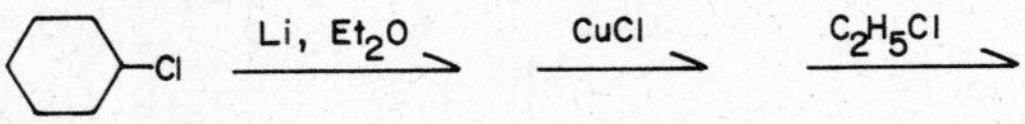

   $$\text{⌬-Cl} \xrightarrow{\text{Li, Et}_2O} \xrightarrow{\text{CuCl}} \xrightarrow{C_2H_5Cl}$$

   (e) an alkyl halide which will yield pure 3-methyl-1-butene on treatment with hot alcoholic potassium hydroxide.

4. (25 pts, 15 min) Draw clear electronic structures (Lewis) including any significant resonance structures if appropriate for (a) nitric acid HNO_3, (b) hydrogen cyanide HCN, (c) diazomethane H_2CN_2, (d) sodium bisulfite $NaHSO_3$, and additionally (e) tell whether the structures below are cations, anions, radicals, or neutral molecules.

   $H_3C-\ddot{O}\substack{H \\ H}$ (I) $H_3C-\ddot{N}\vcentcolon$ (II)

First Examination (D)   Answer Set

1. (a) $Cl_2 \xrightarrow{heat} 2\ Cl\cdot$

   $Cl\cdot + C_2H_6 \longrightarrow C_2H_5\cdot + HCl$

   $2\ C_2H_5\cdot \longrightarrow C_4H_{10}$

   $Cl\cdot + C_4H_{10} \longrightarrow C_4H_9\cdot + HCl$

   or $C_2H_5\cdot + C_4H_{10} \longrightarrow C_4H_9\cdot + C_2H_6$

   $C_4H_9\cdot + Cl_2 \longrightarrow C_4H_9Cl + Cl\cdot$

   Two butyl radicals exist:

   $\underset{I}{CH_3CHCH_2CH_3} \qquad \underset{II}{CH_2CH_2CH_2CH_3}$

   However, I is more stable than II and therefore is formed more readily. Thus 2-chlorobutane will predominate over 1-chlorobutane.

   (b) $CH_3CH_2CH(CH_3)CH_2CH_3 \xrightarrow{Cl_2,\ h\nu}$

   $CH_3CH_2CH(CH_3)CH_2CH_2Cl$ I
   1-chloro-3-methylpentane

   $CH_3CH_2CH(CH_3)CHClCH_3$ II
   2-chloro-3-methylpentane

   $CH_3CH_2CCl(CH_3)CH_2CH_3$ III
   3-chloro-3-methylpentane

   $CH_3CH_2CH(CH_2Cl)CH_2CH_3$ IV
   3-(chloromethyl)pentane

   Note: There are only four monochlorinated derivatives. Any other structures one may try to draw are equivalent to one of the above.
   The remainder of the problem (the numerical part) is solved by following the procedure described in the solution to problem 1 in First Examination (C).

   Relative amounts of products
   I   6 x 1.0 = 6.0
   II   4 x 3.8 = 15.2
   III 1 x 5.0 = 5.0
   IV   3 x 1.0 = 3.0

   Thus

   %I   = 6/29.2 x 100   = 20.5%
   %II = 15.2/29.2 x 100 = 52.1%
   %III = 5.0/29.2 x 100 = 17.1%
   %IV = 3.0/29.2 x 100 = 10.3%

2. (a) Homolytic: $H_2S_2 \longrightarrow 2\ HS\cdot$ (radicals)

   Heterolytic: $H_2S_2 \longrightarrow HS^+ + HS^-$ (cation and anion)

   (b) E_{act} for $Br_2 \longrightarrow 2\ Br\cdot$ is 46 kcal/mole

   E_{act} for $2\ I\cdot \longrightarrow I_2$ is zero

   F_2 reacts most rapidly with CH_4

19

(c)

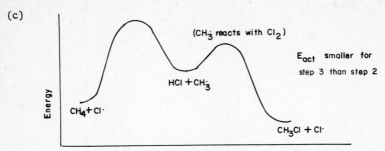

3. (a) 2,2,4,4-tetramethylpentane $(CH_3)_3CCH_2C(CH_3)_3$

   (b) 2,2,3,3-tetramethylbutane $(CH_3)_3CC(CH_3)_3$

   (c) 3-methyl-2-butene $(CH_3)_2C=CHCH_3$

   (d) 1-ethylcyclohexane ⬡—CH_2CH_3

   (e) 3-methyl-1-chlorobutane $(CH_3)_2CHCH_2CH_2Cl$

4. (a) H-Ö-N⁺(=Ö:)(Ö:⁻) ⟷ H-Ö-N⁺(Ö:)(=Ö:)

   (b) H-C≡N:

   (c) H₂C=N⁺=N:⁻ ⟷ H₂C⁻-N⁺≡N:

   (d) Na⁺ [:Ö=S(=Ö:)(Ö-H) ⟷ :Ö⁻-S(=Ö:)(Ö-H)]

   Note that a structure with sodium covalently bound is wrong.

   (e) H₃C-Ö⁺(H)(H)   is a positive ion

   H₃C-N̈   is neutral

The charge on a species is given by the sum of the formal charges on all atoms; a radical has at least one unpaired electron.

First Examination (E) One Hour

1. How are the sulfur and oxygen atoms held together in sulfate (SO_4^{2-}) ions?
   (a) ionic bonds
   (b) covalent bonds
   (c) dipole attractions
   (d) van der Waals forces
   (e) hydrogen bonds

2. Methyl chloride, CH_3Cl, can be made by reaction of Cl_2 with CH_4 in the presence of light or heat. Other di-, tri-, and tetrasubstituted substances may be produced as by-products. What is the best way of limiting the reaction to the production of CH_3Cl?
   (a) by using 1 mole of CH_4 and 1 mole of Cl_2
   (b) by using 1 mole of CH_4 and a large excess of Cl_2
   (c) by using 1 mole of Cl_2 and a large excess of CH_4
   (d) by using 4 moles of Cl_2 and 1 mole of CH_4
   (e) by adding tetramethyl lead as a catalyst to an equimolar mixture of Cl_2 and CH_4

3. What are the hybridization schemes for the carbon and oxygen atoms in CO_2?
   (a) C sp^2, O sp^2
   (b) C sp, O sp^2
   (c) C sp, O sp
   (d) C sp^2, O sp^3
   (e) none of the above

4. Which of the following is the strongest acid?
   (a) HI
   (b) HBr
   (c) HCl
   (d) HF
   (e) H_2Se

5. How can the following equation best be completed to describe one of the chain propagating steps in the reaction of methane with chlorine?
   $CH_3\cdot + Cl_2 \longrightarrow$
   (a) $CH_3Cl + Cl^-$
   (b) $CH_3 + Cl_2\cdot$
   (c) $CH_3Cl + Cl\cdot$
   (d) $HCl + CH_2Cl\cdot$
   (e) $CH_3 + 2\,Cl\cdot$

6. Give the number of primary, secondary, and tertiary hydrogens in 2,3,4-trimethylpentane.
   (a) 1°, 15; 2°, 0; 3°, 3
   (b) 1°, 12; 2°, 3; 3°, 3
   (c) 1°, 12; 2°, 0; 3°, 4
   (d) 1°, 12; 2°, 6; 3°, 0
   (e) none of the above is correct

7. Considering again 2,3,4-trimethylpentane, how many <u>different</u> monochlorinated derivatives can be made by reaction with chlorine in the presence of light (count any enantiomeric pairs as one product)?
   (a) 2
   (b) 3
   (c) 4
   (d) 5
   (e) 6

8. If the relative reactivities of 1°, 2°, and 3° hydrogen atoms toward Cl· are 1.0:3.8:5.0, what percentage by weight of the total yield of monochlorinated products of 2,3,4-trimethylpentane will be accounted for by the most abundant of these products?
   (a) 50%
   (b) 40%
   (c) 83.33%
   (d) 66.66%
   (e) none of the above.

9. In two separate experiments, identical quantities of propane (constituting an excess in each case) were treated with identical molar quantities of Cl_2 and Br_2 under identical conditions for identical periods of time.

   $CH_3CH_2CH_3 \xrightarrow[h\nu]{X_2} CH_3CHCH_3 \quad + \quad CH_3CH_2CH_2X$
   $\phantom{CH_3CH_2CH_3 \xrightarrow[h\nu]{X_2} CH_3CH}|$
   $\phantom{CH_3CH_2CH_3 \xrightarrow[h\nu]{X_2} CH_3CH}X$

   x moles isolated   y moles isolated

   Experiment 1; X = Cl   Experiment 2; X = Br
   For which experiment would (i) x + y, and (ii) x/y be greater?
   (a) Expt. 1, Expt. 1
   (b) Expt. 1, Expt. 2
   (c) Expt. 2, Expt. 1
   (d) Expt. 2, Expt. 2
   (e) same in each experiment

10. If the Fischer projection below is to represent (2S,3S)-2,3-dichloropentane, the identities of groups 1-4 must be as follows:
    (a) 1 = Cl, 2 = Et, 3 = Cl, 4 = CH_3
    (b) 1 = Et, 2 = Cl, 3 = Cl, 4 = CH_3
    (c) 1 = Et, 2 = Cl, 3 = CH_3, 4 = Cl
    (d) 1 = Cl, 2 = Et, 3 = CH_3, 4 = Cl
    (e) 1 = Cl, 2 = CH_3, 3 = Cl, 4 = CH_3

11. What is the correct IUPAC name for the alkyl halide which will yield pure 4-ethyl-1-hexene on treatment with KOH/EtOH? If no alkyl halide will yield pure alkene, choose answer (e).
    (a) 1-chloro-3-ethylhexane

(b) 2-chloro-4-ethylhexane
(c) 1-chloro-4-ethylhexane
(d) 6-chloro-3-ethylhexane
(e) no alkyl halide can yield pure alkene under these conditions

12. How many different alkenes have the molecular formula C_5H_{10}?
    (a) 4
    (b) 5
    (c) 6
    (d) 8
    (e) none of the above

13. Consider nitromethane, CH_3NO_2, for which two significant resonance structures may be drawn. The formal charge on the nitrogen atom in these two structures is
    (a) +1, +1
    (b) 0, 0
    (c) 0, -1
    (d) +1, -1
    (e) none of these

14. Give the name of the product resulting from the following sequence:
    n-propyl bromide + (1) Mg/Et$_2$O; (2) H$_2$O
    (a) n-propanol
    (b) n-propane
    (c) isopropanol
    (d) propene
    (e) hexane

15. Repeat question number 14 for:
    2-chlorobutane + (1) Li, CuI; (2) 1-bromopentane
    (a) nonane
    (b) 1-methyl-2-ethylcyclohexane
    (c) 3-methyloctane
    (d) 1-ethylcycloheptane
    (e) none of the above

16. From each pair, pick the compound that undergoes dehydration most readily.
    (i) A 1-hexanol or B 2-hexanol
    (ii) C 1-methylcyclopentanol or D 3-methylcyclopentanol
    (iii) E 2,2,3-trimethyl-2-hexanol or F 2,2,3-trimethyl-3-hexanol
    (a) A, C, E
    (b) A, C, F
    (c) B, D, E
    (d) B, D, F
    (e) none of the above is correct

17. From each pair pick the compound with the higher boiling point.
    (i) A cis-1,2-dichloroethene or B trans-1,2-dichloroethene
    (ii) C n-pentane or D neopentane

(iii) E methane or F methyl chloride
(a) A, C, E
(b) A, D, E
(c) B, D, F
(d) A, C, F
(e) none of the above is correct

18. The reaction below yields a product (C_8H_{14}) which contains no cyclopentane ring. What is the likely product?

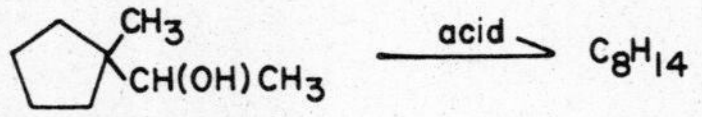

    cyclopentane-CH(OH)CH₃ with CH₃ substituent $\xrightarrow{acid}$ C_8H_{14}

    (a) cyclooctene
    (b) 1,2-dimethylcyclohexene
    (c) 3,3-dimethylcyclohexene
    (d) 1,2,3,4-tetramethylcyclobutene
    (e) 1-octyne

19. Place in order of stability (most stable first):

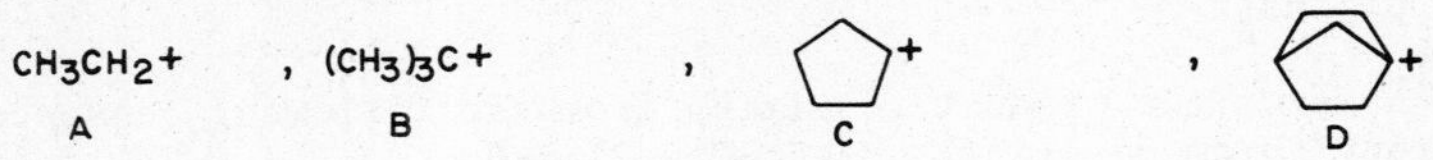

    $CH_3CH_2^+$, $(CH_3)_3C^+$, cyclopentyl cation , cyclohexyl cation
       A   B   C   D

    (a) D, B, C, A
    (b) B, C, A, D
    (c) B, A, C, D
    (d) D, B, A, C
    (e) none of the above is correct

20. Give the major product resulting from the following sequence:
    3-chloro-2,2-dimethylbutane + (1) KOH, EtOH; (2) H_2, Pt
    (a) 2,3-dimethylbutane
    (b) 1-butyne
    (c) 2-butyne
    (d) 2,2-dimethylbutane
    (e) isohexane

First Examination (E)   Answer Set

1.(b) 2.(c) 3.(c) 4.(a) 5.(c) 6.(a) 7.(c) 8.(b) 9.(b) 10.(a) 11.(c) 12.(c)

13.(a) 14.(b) 15.(c) 16.(e) 17.(d) 18.(b) 19.(b) 20.(d)

Second Examination (A)

One Hour

1. (25 pts, 15 min) Complete the following (each part of equal credit):

   (a) 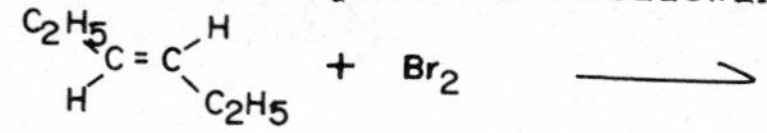 $\underset{H}{\overset{C_2H_5}{>}}C=C\underset{C_2H_5}{\overset{H}{<}}$ + Br_2 ⟶ (give Fischer projection)

   (b) $C_2H_5C\equiv CH$ + C_2H_5MgBr ⟶

   (c) 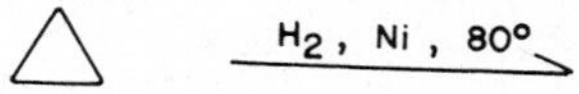 △ $\xrightarrow{H_2,\ Ni,\ 80°}$

   (d) $(CH_3)_2C=CH_2$ + $CHCl_3$, KOH $\xrightarrow{(CH_3)_3COH}$

   (e) $C_2H_5C\equiv CH$ + HBr (excess) ⟶

2. (17 pts, 10 min) A hydrocarbon $C_{11}H_{20}$ absorbs two molar equivalents of hydrogen when shaken with hydrogen gas in the presence of a platinum catalyst. Oxidation of the original hydrocarbon with hot aqueous permanganate yields a mixture containing methyl ethyl ketone, acetone, and succinic acid ($HOOCCH_2COOH$). Deduce the structure of the hydrocarbon and show your reasoning.

3. (25 pts, 15 min) Substances I and II react giving a mixture of III and IV. At very low temperatures, more III than IV is obtained, whereas at higher temperatures more IV is obtained. It is also known that prolonged heating of either pure III or pure IV always results in a mixture of III and IV, and, furthermore, the relative amounts of III and IV in this mixture are always the same, with IV dominating. Draw a reaction progress diagram, and explain the reactions above. Do not assume any specific structures for I and II; the answer should be a general one.

4. (18 pts, 10 min) What is the difference in energy between structure I (shown below) and its other chair-type conformational isomer (not shown, but must be drawn as part of the answer)?

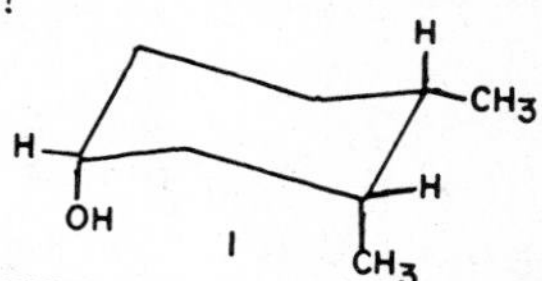

   Information needed:
   methyl/methyl gauche interaction energy   0.9 kcal/mole
   hydrogen/methyl 1,3 diaxial interaction energy   0.9 kcal/mole
   hydrogen/hydrogen 1,3 diaxial interaction energy 0.25 kcal/mole
   methyl/hydroxyl 1,3 diaxial interaction energy   2.3 kcal/mole

   Indicate whether I is more or less stable than its conformational isomer.

5. (15 pts, 10 min) Suggest methods for the following synthetic conversions:

(a) $H_2C_2 \longrightarrow CH_3\underset{O}{\overset{\parallel}{C}}C_2H_5$

(b) cyclohexanol $\longrightarrow$ 1-bromo-2,2-dideuteriocyclohexane

In each case one may use any additional reagents, as long as one begins with the specified starting material and ends up with the specified product.

Second Examination (A)   Answer Set

1. (a) C_2H_5 $\,\,\,$H
       C=C $\xrightarrow{Br_2}$
       H $\,\,\,\,\,$ C_2H_5

   (product: Br—CH(C$_2$H$_5$)—CH(C$_2$H$_5$)—Br, with Br and H anti)

   This reaction involves the intermediate formation of a cyclic bromonium ion which is opened by <u>anti</u> attack of bromide.

   (b) $C_2H_5C\equiv CH + C_2H_5MgBr \longrightarrow C_2H_5C\equiv CMgBr + C_2H_6$

   The acidic acetylenic hydrogen atom is removed by the Grignard reagent.

   (c) △ $\xrightarrow{H_2,\,Ni,\,80°}$ $CH_3CH_2CH_3$

   Formation of an open chain compound relieves the ring strain in the cyclopropane ring.

   (d) $(CH_3)_2C=CH_2$ $\xrightarrow{CHCl_3,\,KOH,\,(CH_3)_3COH}$

   H_3C—◁—Cl
   H_3C $\,\,\,$ Cl

   (cyclopropane with two CH$_3$ on one carbon and two Cl on another)

   This reaction involves the intermediate formation of dichlorocarbene, :CCl$_2$, which reacts with the alkene giving a cyclopropane derivative.

   (e) $H_5C_2C\equiv CH + HBr \longrightarrow H_5C_2CBr=CH_2 \xrightarrow{HBr} H_5C_2CBr_2CH_3$

   Both stages of HBr addition take place with Markownikov orientation.

2. The formation of methyl ethyl ketone and of acetone indicates that the parent hydrocarbon contains the following terminal groups:

   $CH_3\overset{\overset{O}{\|}}{C}C_2H_5$ $\qquad$ $CH_3\overset{\overset{O}{\|}}{C}CH_3$

   The other oxidation product, succinic acid, shows that the "middle" portion of the parent hydrocarbon contains the unit

   $=CHCH_2CH_2CH=$

   Thus the full structure of the hydrocarbon $C_{11}H_{20}$ is

   H_3C $\,\,\,\,\,\,$ H $\,\,\,\,$ H
   C=C $\,\,\,\,\,\,\,\,\,$ C=C(CH$_3$)$_2$ $\qquad$ (E- and Z- isomers are possible)
   H_5C_2 $\,\,\,$ CH$_2$CH$_2$

3. Since the heating of either pure III or pure IV gives the same mixture of III and IV, there must be a slowly established equilibrium

   III ⇌ IV

   Further, since IV is the predominant compound in the equilibrium mixture, it must be the most stable (i.e. possess a smaller amount of potential energy than III). The enhanced amount of III relative to IV produced at low temperatures suggests that the pathway to IV must involve a greater energy of activation. A good rationalization of these observations, then, is that I and II react to give an intermediate X for which competing pathways to III and IV exist, with III being the less stable, but also formed through a lower energy (more stable) transition state. Thus at low temperatures the reaction is kinetically controlled, whereas at higher temperatures it is controlled by thermodynamic factors. The following diagrams summarize these arguments:

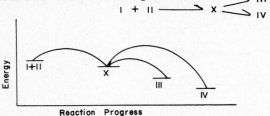

The conversion of X to III and IV is also conveniently represented as:

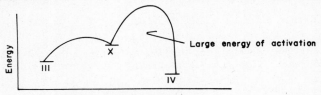

Note that the E_{act} for X going to either III or IV must be much less than to I and II or the reaction would be fully reversible. (Students should recognize that the case of 1,2 versus 1,4 additions to dienes is a specific example of the type of system above.)

4. The conformational isomer of I is shown below as Ia.

A methyl/methyl <u>gauche</u> interaction is present in both I and Ia.

Since this is present in both conformational isomers, one need only consider 1,3 diaxial interactions to compute the energy difference.

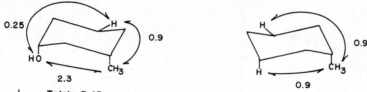

Structure Ia is more stable owing to fewer unfavorable interactions.

5. (a) $H_2C_2 \xrightarrow{NaNH_2} HC_2Na \xrightarrow{CH_3Br} HC{\equiv}CCH_3 \xrightarrow{NaNH_2} NaC{\equiv}CCH_3$

$\xrightarrow{CH_3Br} H_3C\,C{\equiv}CCH_3$

$H_3C\overset{O}{\underset{\|}{C}}C_2H_5 \xleftarrow{H_2O,\ H_2SO_4,\ HgSO_4} H_3C\,C{\equiv}CCH_3$

(b) cyclohexanol $\xrightarrow{acid}$ cyclohexene $\xrightarrow{N\text{-bromosuccinimide}}$ 3-bromocyclohexene $\xrightarrow{D_2,\ catalyst}$ trans-1-bromo-2,3-dideuteriocyclohexane

Second Examination (B)   One Hour

1. (25 pts, 15 min) Complete the following:
   (a) cyclohexene + Br₂ →   Give name and stereochemical drawing for the compound.
   (b) cyclohexene + cold aq. KMnO₄ →   Give name and stereochemical drawing for the product.
   (c) 1-bromo-2-methylcyclohexane (Br, CH₃) + KOH, EtOH →

2. (25 pts, 15 min) Show how the following transformations can be completed with the use of additional organic and inorganic reagents of ones choice.
   (a) acetylene ——→ meso-3,4-dihydroxyhexane
   (b) 3-hexyne ——→ cis-1,2-dimethylcyclopropane
   (c) 2-bromo-2-methylpentane ——→ 3-bromo-2-methylpentane

3. (25 pts, 15 min) Explain each of the following:
   (a) Bromoalkenes are usually converted to alkynes by treatment with KNH_2. 1-Bromocyclohexene does not undergo dehydrohalogenation under these conditions.

   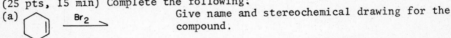

   (b) Using a stereochemical drawing, give the mechanism of the following reaction and thereby deduce the structure of the product.

   (1R,2R)-1,2-dibromo-1,2-diphenylethane —base→ a vinyl bromide $C_{14}H_{11}Br$

   (c) The acidities of acetylene, ethylene, and ethane are different.

4. (25 pts, 15 min) Suggest structures for the following lettered compounds:

   B —acid→ C (C_9H_{14}) —1. O₃; 2. H₂O, Zn or hot aq. KMnO₄→ E (1,3-diacetylcyclopentane)

   (A secondary alcohol)   (The acetyl group is $CH_3\underset{O}{\overset{\parallel}{C}}-$)

Second Examination (B)   Answer Set

1. (a) Bromination of an alkene occurs via <u>anti</u> addition to a cyclic bromonium ion.

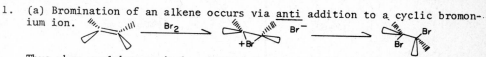

Thus when cyclohexene is brominated the product is <u>trans</u>-1,2-dibromocyclohexane. The most stable conformation of the product is therefore a chair in which both bromine atoms occupy equatorial positions.

(b) Treatment of cyclohexene with cold aq. potassium permanganate results in the formation of <u>cis</u>-cyclohexene-1,2-diol, owing to the <u>syn</u> nature of the addition. Thus one hydroxyl group must occupy an equatorial position and one an axial position.

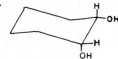

(c) Dehydrohalogenation will take place to give the most stable product.

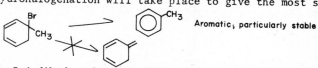

Aromatic; particularly stable

2. (a) <u>Meso</u>-3,4-dihydroxyhexane can be made by the reaction of performic acid (HCO_3H) with the appropriate <u>trans</u>-alkene, or by treatment of the corresponding <u>cis</u>-alkene with cold aqueous potassium permanganate. Thus one needs to make either compound I or II.

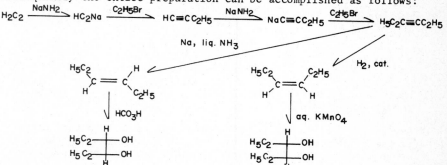

Consequently the entire preparation can be accomplished as follows:

31

(b) <u>Cis</u>-1,2-diethylcyclopropane can be made by addition of singlet methylene to <u>cis</u>-3-hexene. Thus the following sequence of steps is necessary:

$$H_5C_2C{\equiv}CC_2H_5 \xrightarrow{H_2,\ cat.} \underset{H_5C_2}{\overset{H}{}}C=C\underset{C_2H_5}{\overset{H}{}} \xrightarrow{CH_2N_2,\ h\nu\ in\ liquid\ olefin} \text{(cis-1,2-diethylcyclopropane)}$$

(c) The required product is obtained by anti-Markownikov addition to 2-methylpentene. Thus the following procedure suggests itself:
   (i) dehydrohalogentation of the starting material to give 2-methylpentene
   (ii) reaction of 2-methylpentene with HBr in the presence of peroxides

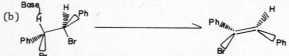

3. (a) The product is too strained to be made. In an alkyne the hybridization for the carbon atoms involved in the triple bond is of the <u>sp</u> type. This necessitates linear geometry requiring the following shape:

$$\underset{CH_2-C{\equiv}C-CH_2}{\overset{CH_2-CH_2}{\diagup\hspace{2em}\diagdown}}$$

In fact, the two methylenes could not bridge across the four linear carbon atoms. Any attempt by the molecule to distort from the geometry described above would result in weakened bonds, i.e. strain.

(b) [structure showing elimination with Base, H, Ph, Br substituents → alkene product with Ph, H, Br]

The eliminated H and Br must bear an <u>anti</u> relationship to one another.

(c) The simplest way to compare the acidities of the hydrocarbons is to look at the relative basicities of their conjugate bases.

$H-C{\equiv}C{:}^-$ Electrons in <u>sp</u> hybrid orbital, 50% s-character, relatively stable

$H_2C=C{:}^-\overset{H}{}$ Electrons in $\underline{sp}^2$ hybrid orbital, 33% s-character, more reactive

$H_3C-C{:}^-\overset{H}{\underset{H}{}}$ Electrons in $\underline{sp}^3$ hybrid orbital, 25% s-character, most reactive of the three anions

Acetylene gives rise to the weakest conjugate base; acetylene is therefore the strongest acid.

Students are often puzzled by an apparent anomaly in this context. Looking at the hydrocarbons themselves, and using arguments similar to the above about hybridization schemes and relative <u>s</u>-character, one can predict (correctly) that the C-H bond in acetylene is stronger (shorter) than that in the other two hydrocarbons. Therefore, shouldn't this make acetylene the <u>weakest</u> acid?

The apparent difficulty is removed if one realizes that acidity is a composite of many factors beside bond strength. In fact, one can look at the process of anion formation in a number of steps:

<u>Step 1</u>. Fission of a C-H bond

$$R-H \longrightarrow R\cdot + H\cdot \qquad \Delta H = R\text{-}H \text{ bond energy}$$

<u>Step 2</u>. Creation of an ion pair

$$R\cdot + e^- \longrightarrow R^- \qquad \Delta H = \text{electron affinity of } R\cdot$$

$$H\cdot \longrightarrow H^+ + e^- \qquad \Delta H = \text{ionization energy of } H\cdot$$

<u>Step 3</u>. Solvation of the ions

$$R^- + H_2O \longrightarrow R^-(aq) \qquad \Delta H = \text{solvation energy of } R^-$$

$$H^+ + H_2O \longrightarrow H^+(aq) \qquad \Delta H = \text{solvation energy of } H^+$$

In step 2, the relative electron affinities of different hydrocarbon radicals govern the ease of formation of R^-. Comparing acetylide ion with the other species one sees that in step 2 an added electron goes into an orbital with more <u>s</u>-character for the acetylenic species. This results in greater stability and easier formation. This extra stability more than offsets the greater amount of energy needed to break the acetylene C-H bond in step 1. Thus acetylene is still predicted to be the strongest acid although now the argument is more subtle. Rationalization in terms of the conjugate bases is generally preferred since it avoids such subtleties.

4. This problem is solved fairly simply by working backwards from the specified structure (E), 1,3-diacetylcyclopentane.

Since this was formed by an ozonolysis reaction on (C), and because (C) contains the same number of carbon atoms as (E), it is clear that (C) must be a bicyclic molecule.

The fact that permanganate degradation of (C) gives the same product as ozone degradation further confirms that (C) is a symmetrical olefin of the structure indicated. Compound (C) is formed by dehydration of a 2° alcohol. This shows that a methyl group must have migrated in the dehydration step, since any alcohol which could dehydrate to (C) and at the same time have the same carbon skeleton as (C) would be 3°. Thus the conversion (B)-(C) can be envisioned as follows:

33

Note that the following are alternative, and equally acceptable representations of (B):

Second Examination (C)   One Hour

1. (25 pts, 26 min) A compound A (C_5H_9Br) gives formaldehyde and another aldehyde B when ozonized. When A is treated with HBr a compound C ($C_5H_{10}Br_2$) is produced. When C is heated with zinc it gives a substance D (C_5H_{10}) as one of the products. Compound D is also produced by the following sequence:

   1-bromo-1-bromomethylcyclobutane $\xrightarrow{\text{zinc}}$ E(C_5H_8) $\xrightarrow{H_2,\text{cat.}}$ D(C_5H_{10})

   Give structures for the compounds (A)-(E).

2. (18 pts, 12 min) Consider the following structures:

   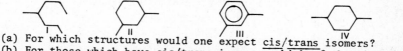

   (a) For which structures would one expect cis/trans isomers?
   (b) For those which have cis/trans isomers, which of the cis or the trans will be more stable? Explain carefully using stereochemical or conformational drawings.

3. (18 pts, 12 min) How could the following conversions be effected?
   (a)
   (b)
   (c) $(CH_3)_2C=C(CH_3)_2 \longrightarrow (CH_3)_2CHC(CH_3)CH_2Br$
                                                       |
                                                       Br

4. (9 pts, 6 min) Explain the fact that cyclopentadiene is "unusually" acidic for a hydrocarbon species.

5. (12 pts, 8 min) Draw Fischer projection formulas for the missing compounds in the following reactions. (Note: each question mark may correspond to more than one compound. Each answer should include all possibilities.)

   (a) ? (C_7H_{14}) $\xrightarrow{KMnO_4, H_2O}$

   CH$_3$
   H—|—OH
   H$_3$C—|—OH
   CH$_2$CH$_2$CH$_3$

   (b) (R)-CH$_3$CH$_2$CHCH=CH$_2$
                 |
                CH$_3$

   $\xrightarrow{\text{HBr, peroxide}}$ ($C_6H_{13}Br$) ?

   $\xrightarrow{\text{HBr, no peroxide}}$ ($C_6H_{13}Br$) ?

6. (18 pts, 12 min) Supply structures for the lettered compounds.

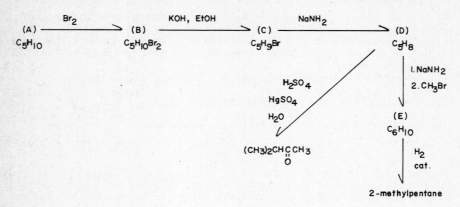

Second Examination (C)   Answer Set

1. The structure of D is established from the information provided in the last part of the problem

   [cyclobutane-CH₂Br with Br] —Zn→ (E) [methylenecyclobutane] —H₂, cat.→ (D) [methylcyclobutane]

   Now, since A gives formaldehyde on ozonolysis, it must be a terminal olefin. Reaction of A with HBr will proceed with Markownikov orientation in the absence of peroxides, so a partial structure for C can be written:

   ~~~C(Br)(CH₃)

   Treatment of C with Zn results in the loss of two bromine atoms, producing D. Thus a ring forming reaction has occurred with the methyl group in C becoming the methyl group in D. The position of the second bromine atom in C, and hence the single bromine atom in A, can be deduced from the size of the ring (4 carbons). Thus the two bromine atoms in C are separated by four carbon atoms. In summary:

   (A) [Br-CH₂-CH₂-CH₂-CH=CH₂] —HBr→ (C) [Br-CH₂-CH₂-CH₂-CH(Br)-CH₃] → [methylcyclobutane] ← (D) ← [cyclobutyl-CH₂Br]

   ↓ O₃, work up

   (B) [Br-CH₂-CH₂-CH₂-CHO]

2. Only structures II and IV can represent both cis and trans isomers. Structure I is that of an open chain alkane, and free rotation about the single bonds precludes the possibility of separate cis and trans isomers. Structure III is that of a benzene derivative. The aromatic ring system is planar and both substituents lie in the plane of the ring. No cis/trans isomers are possible.

   Conformational drawings for the cis and trans isomers relevant to structures II and IV are as follows:

   Structure II

   [chair conformation, cis-]   [chair conformation, trans-]

   The cis isomer is more stable because there are no undesirable 1,3-diaxial interactions.

   Structure IV

   [chair conformation, cis-]   [chair conformation, trans-]

In this case the trans isomer is the more stable one because with both methyl groups in equatorial positions there are no 1,3-diaxial methyl-hydrogen interactions.

3. (a) [cyclohexyl-CH=CH-CH₃] $\xrightarrow{Na}{liq.\ NH_3}$ [cyclohexyl-C≡CH] $\xrightarrow{KOH}{heat}$ [cyclohexyl-CHBr-CHBr-] $\xleftarrow{Br_2}$ [cyclohexyl-CH=CH-] (cis)

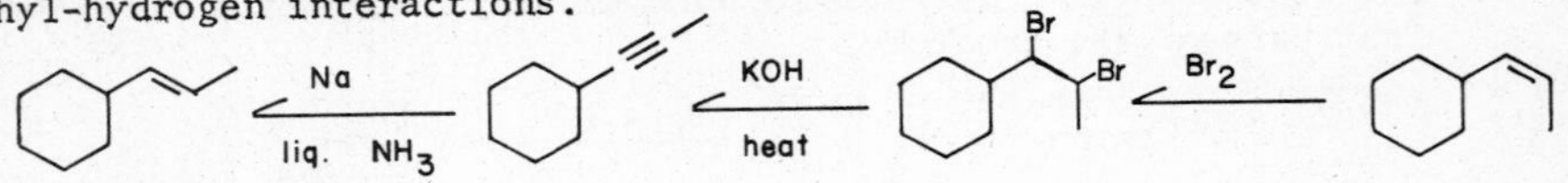

In this part of the question one sees that conversion of a cis alkene to a trans alkene can be accomplished by going through an intermediate alkyne. (How could one change a trans alkene to a cis alkene?)

(b) [cyclopentane with H, OCH₃, Br, H] $\xleftarrow{Br_2}{CH_3OH}$ [cyclopentene] $\xleftarrow{KOH}{EtOH}$ [cyclopentane with Br] $\xleftarrow{Br_2, heat}{(free\ radical)}$ [cyclopentane]

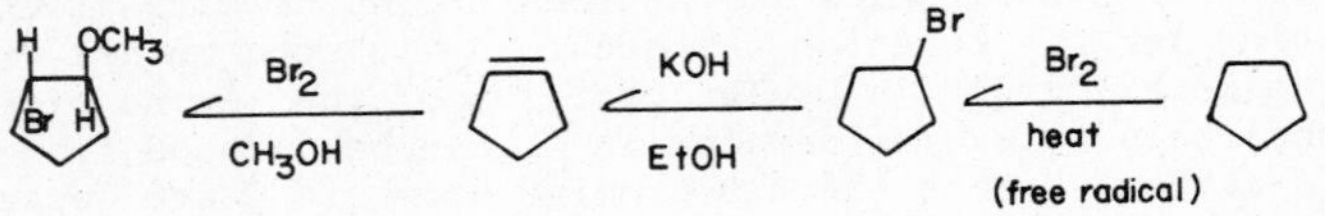

The final step in this sequence involves the attack of methanol on the bromonium ion formed by attack of bromine on cyclopentene.

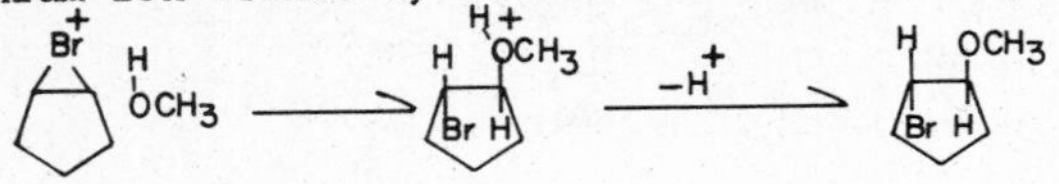

[Br⁺ cyclopentane with H, OCH₃] $\rightarrow$ [H⁺ H OCH₃ cyclopentane with Br, H] $\xrightarrow{-H^+}$ [H OCH₃ cyclopentane with Br, H]

(c) $(CH_3)_2CHC(CH_3)CH_2Br$ with Br $\xleftarrow{Br_2}$ $(CH_3)_2CHC=CH_2$ with CH₃ $\xleftarrow{KOH,\ EtOH}$ $(CH_3)_2CHCHCH_2Br$ with CH₃

$(CH_3)_2C=C(CH_3)_2$ $\xrightarrow{NBS}$ $(CH_3)_2C=CCH_2Br$ with CH₃ $\xrightarrow{H_2,\ cat.}$

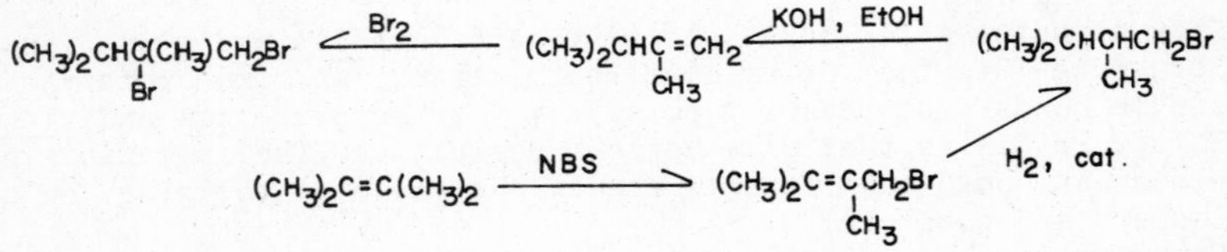

4. Charge delocalization results in a very significant stabilization for the cyclopentadienyl anion; an aromatic system is generated, one of a cyclic π system bearing (4n+2)π electrons. This is illustrated by the resonance structures shown below.

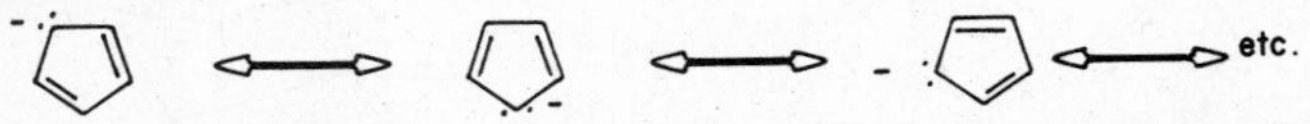

[cyclopentadienyl anion resonance structures] ⟷ [ ] ⟷ [ ] ⟷ etc.

Thus, since the conjugate base of cyclopentadiene is especially stable, it is relatively easy for cyclopentadiene to lose a proton. Accordingly, it is a relatively acidic compound.

5. (a) Since the use of cold aq. $KMnO_4$ results in the syn addition of two hydroxyl groups to an alkene, one must keep in mind that both hydroxyls add to the same side of the molecule. One should work backwards, stepwise, from the Fischer projection given to the starting alkene.

[Structural diagram: Fischer-like projection with CH₃ top, H—OH, H₃C—OH, C₃H₇ bottom; equivalence to 3D structure with HO, OH, H₃C, C₃H₇, CH₃]

[Reaction scheme: alkene with H₃C and H₇C₃ groups + cold aq. KMnO₄ → diol with HO, OH, H₃C, C₃H₇, CH₃]

Since both hydroxyls are on the same side, the position of the alkyl groups must reflect their position in the starting alkene.

(b) First part – peroxide-initiated reaction

[Scheme: alkene →(Br·) alkyl radical with Br →(HBr) alkyl bromide with Br]

No new chiral carbon atom is generated in this reaction.

Second part – carbonium ion mechanism

$$\begin{array}{c} CH=CH_2 \\ H{-}{-}CH_3 \\ C_2H_5 \end{array} \longrightarrow \begin{array}{c} CH_3 \\ H{-}{-}Br \\ H{-}{-}CH_3 \\ C_2H_5 \end{array} + \begin{array}{c} CH_3 \\ Br{-}{-}H \\ H{-}{-}CH_3 \\ C_2H_5 \end{array}$$

A new chiral carbon center is generated; products are diastereomers. The alkene abstracts a proton from the HBr, forming a carbonium ion that is then attacked by bromide ion to give the products.

[Scheme: H₅C₂\CHCH=CH₂ /H₃C →(HBr) planar carbonium ion H₅C₂\CHCHCH₃ + Br⁻ ; →(HBr, crossed out) H₅C₂\CHCH₂CH₂ + Br⁻]

6. The structure of compound $\underline{D}$ can be deduced from that of 3-methyl-2-butanone which, one is told, is formed by treatment of $\underline{D}$ with HgSO₄/H₂SO₄.

$$(CH_3)_2CH\overset{O}{C}CH_3 \xleftarrow{HgSO_4,\ H_2SO_4,\ H_2O} (CH_3)_2CHC\equiv CH$$
$$\hspace{8cm} \underline{D}$$

The fact that 2-methylpentane is produced by the reduction of the product obtained from the sodium salt of $\underline{D}$ and methyl iodide further confirms the skeleton of $\underline{D}$.

$$(CH_3)_2CHC\equiv CH \xrightarrow[2.\ CH_3Br]{1.\ NaNH_2} (CH_3)_2CHC\equiv CCH_3 \xrightarrow[catalyst]{H_2} (CH_3)_2CHC_3H_7$$
$$\hspace{6cm} E$$

The structures of the remaining lettered compounds are then readily deduced.

(CH₃)₂CHCH=CH₂   (CH₃)₂CHCH(Br)CH₂Br   (CH₃)₂CHCH=CHBr
    A   B   C

Second Examination (D)   One Hour

1. (25 pts, 15 min) Complete the following reactions:

   (a) 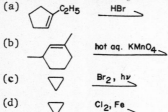 cyclopentene with C₂H₅ substituent, HBr

   (b) methylcyclohexene with exocyclic methyl, hot aq. KMnO₄

   (c) cyclopropene, Br₂, hν

   (d) cyclopropene, Cl₂, Fe

   (e) 1,3-cyclohexadiene, HCl (1 mole)

2. (25 pts, 15 min)
   (a) Place in order of heats of combustion per CH_2 unit (most exothermic first) the following:
       A. cyclopropane B. cyclobutane C. cyclohexane D. propane
       If any two are the same, so indicate.
   (b) Repeat (a) for heats of hydrogenation per double bond for:
       A. 1,4-pentadiene B. 1,3-pentadiene
   (c) Repeat (b) for:
       A. pentene B. 1,3-pentadiene C. 1,3-butadiene
   (d) How many rings (if any) are present in a compound $C_{42}H_{76}$ if it yields another compound $C_{42}H_{80}$ upon catalytic hydrogenation?
   (e) Which of the following are chiral? Explain.
       A. <u>trans</u>-1,3-dimethyl-<u>trans</u>-2,4-dibromocyclobutane
       B. <u>trans</u>-dimethylcyclopropane
       C. <u>cis</u>-dimethylcyclopropane
   Show structures as part of the answer.

3. (25 pts, 15 min) Using 0.9 kcal/mole for each 1-3 CH_3-H diaxial interaction with 3.4 kcal/mole for each 1-3 CH_3-CH_3 diaxial interaction and 0.9 kcal/mole for each butane gauche interaction, compute the energy difference between the two chair forms of the following:

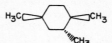

   Show both chair forms in the answer.

4. (25 pts, 15 min) Deduce the structures of compounds <u>A-H</u>.

40

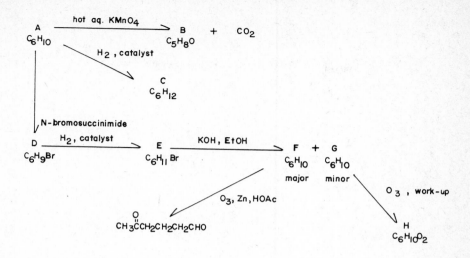

Second Examination (D)   Answer Set

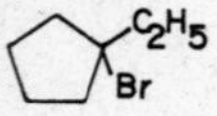

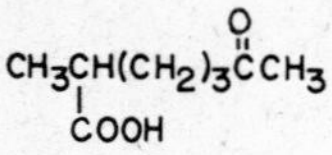

1. [Structures A–E: A = bromoethylcyclopentane (Br, C₂H₅); B = CH₃CH(CH₂)₃CCH₃ with COOH; C = bromocyclopropane; D = ClCH₂CH₂CH₂Cl; E = chlorocyclohexene]

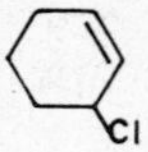

2. (a) A, B, C D
   (b) A, B
   (c) A, C, B
   (d) Three rings
   (e)

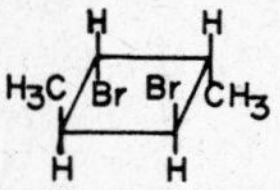

   [Structure: cyclobutane with H₃C/Br and Br/CH₃ substituents]
   Achiral; it has a center of symmetry.

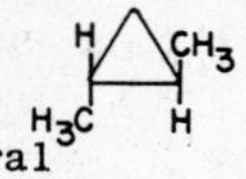

   [Structure: cyclopropane with H₃C and CH₃]
   Chiral

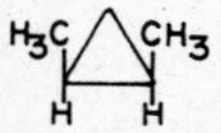

   [Structure: cyclopropane with H₃C and CH₃]
   Achiral; it has a plane of symmetry.

3. [Two chair conformations of a methylated cyclohexane shown]

   The two chair conformations are shown above. The one on the left is more stable by 0.9 kcal/mole.

4. This problem should be approached by working backwards from the given structure.

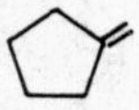

   [Structures A–H]
   A: methylenecyclopentane
   B: cyclopentanone
   C: methylcyclopentane
   D: 1-bromo-2-methylenecyclopentane
   E: 1-bromo-2-methylcyclopentane
   F: 1-methylcyclopentene
   G: 3-methylcyclopentene
   H: OHC-CH₂-CH₂-CH(CH₃)-CHO

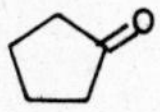

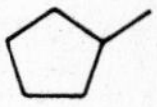

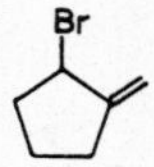

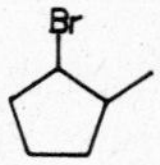

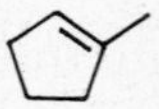

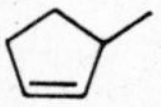

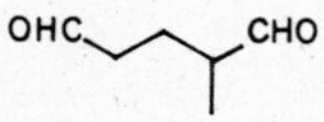

Second Examination (E)   One Hour

1. 1-Butene is subjected to the following sequence of reactions: (1) allylic bromination to give C_4H_7Br, and (2) treatment of this product with dilute aq. permanganate. If a distillation is performed at the end of the sequence, how many fractions will be obtained?
   (a) 1
   (b) 2
   (c) 3
   (d) 4
   (e) 8

2. A racemic mixture of (2R,3R)- and (2S,3S)-2,3-butanediol is obtained by the reaction of
   (a) cis-2-butene with $(BH_3)_2$, HO_2^-
   (b) cis-2-butene with dil. aq. permanganate
   (c) trans-2-butene with $(BH_3)_2$, HO_2^-
   (d) trans-2-butene with dil. aq. permanganate
   (e) a mixture of cis- and trans-2-butene with dil. aq. permanganate

3. The number of tertiary hydrogen atoms in bicyclo[2.2.1]heptane is
   (a) 0
   (b) 1
   (c) 2
   (d) 3
   (e) 4

4. Give the product of the following sequence:
   cyclohexanol + (1) acid, heat, (2) N-bromosuccinimide, (3) KOH, EtOH
   (a) 3-cyclohexenol
   (b) cyclohexyne
   (c) 1,2-cyclohexadiene
   (d) 1,3-cyclohexadiene
   (e) 2-bromocyclohexanol

5. Consider the following bonds:
   A. The C-C bond in ethylene
   B. The C-C bond in ethane
   C. The C_1-C_2 bond in 1,3-butadiene
   D. The C_2-C_3 bond in 1,3-butadiene
   The correct order in terms of increasing bond length (longest last) is
   (a) ACDB
   (b) ADCB
   (c) ABCD
   (d) ACBD
   (e) CADB

6. How many stereoisomers are there for $HOCH_2CH(OH)CH(OH)CH(OH)CH_3$?
   (a) 4
   (b) 6

(c) 8
   (d) 16
   (e) None of the above is correct.

7. 2-Methylpropene + Cl_2/H_2O yields
   (a) $(CH_3)_2CClCH_2OH$
   (b) $(CH_3)_2COHCH_2Cl$
   (c) 1,2-epoxy-2-methylpropane
   (d) $CH_3CH(CH_2OH)CH_2Cl$
   (e) 2,3-epoxybutane

8. When cis-2-butene is treated with bromine in the dark there is formed
   (a) meso-2,3-dibromobutane
   (b) a mixture in which meso-2,3-dibromobutane is the major constituent
   (c) a racemic mixture of d- and l- 2,3-dibromobutane
   (d) a mixture of d-2,3-dibromobutane and meso-2,3-dibromobutane
   (e) nothing; starting materials are recovered

9. A mixture of 2-decanone and 3-decanone is produced by the reaction of compound A, $C_{10}H_{18}$, with water in the presence of $HgSO_4$ and H_2SO_4. A is most probably
   (a) 1-decyne
   (b) 2-decyne
   (c) 3-decyne
   (d) 3-methylcyclononyne
   (e) 3,8-dimethylcyclooctyne

10. In order to convert 3-methylbutyne to 1,2-dibromo-3-methylbutane the following reaction sequence should be followed:
    (a) bromine, then HBr
    (b) HBr (two moles)
    (c) HBr (one mole), then HBr/peroxide
    (d) HBr, then bromine
    (e) HBr (two moles) in presence of peroxide

11. An alkene $C_{10}H_{20}$ yields a mixture of 2-pentanone and 3-pentanone when treated with ozone followed by treatment with Zn/acetic acid. The alkene is possibly
    (a) 1-methyl-2-propylcyclopentene
    (b) 1,2-dimethylcyclopentene
    (c) 3-pentene
    (d) 3-ethyl-4-methyl-3-heptene
    (e) (E)-3,4-dimethyl-3-hexene

12. What is the formal charge on the carbon atom of carbene methylene?
    (a) -1
    (b) 0
    (c) +1
    (d) +2
    (e) -2

13. What is the product of the following sequence?
    cyclohexanol + (1) acid, heat, (2) formic acid, hydrogen peroxide
    (a) 1,5-hexanedioic acid
    (b) cyclohexanone
    (c) 3-cyclohexenol
    (d) cis-1,2-cyclohexanediol
    (e) trans-1,2-cyclohexanediol

14. How many π electrons are there in A and B?

    (a) 4,2
    (b) 6,2
    (c) 6,4
    (d) 4,4
    (e) none of the above is correct

15. Which of the molecules A-D are aromatic?

    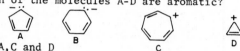

    (a) A,C and D
    (b) B,C and D
    (c) C and D
    (d) A and D
    (e) all of the molecules

16. Arrange in order of heat of hydrogenation per double bond the following compounds (most exothermic should be first):
    A. cyclohexene
    B. benzene
    C. 1,3,5-hexatriene
    D. 1,3-cyclohexadiene
    (a) B,C,D,A
    (b) B,D,C,A
    (c) C,A,D,B
    (d) A,D,C,B
    (e) none of the above is correct

17. In the most stable conformation of A, how many methyl groups are in equatorial positions?
    (a) 0
    (b) 1
    (c) 2
    (d) 3

18. On the basis of 0.9 kcal/mole per 1,3-CH_3-H diaxial interaction, what is the energy difference between the two chair conformations of trans-1,4-dimethylcyclohexane?

(a) 0.9 kcal/mole
(b) 1.8 kcal/mole
(c) 2.7 kcal/mole
(d) 3.6 kcal/mole
(e) 4.5 kcal/mole

19. For each pair, pick the more stable molecule.
    (i) A singlet methylene; B triplet methylene
    (ii) C trans-1,2-dimethylcyclohexane; D cis-1,2-dimethylcyclohexane
    (iii) E trans-1,3-dimethylcyclohexane; F cis-1,3-dimethylcyclohexane
(a) B,C,E
(b) A,D,E
(c) A,C,F
(d) B,C,F
(e) none of the above is correct

20. Consider the following compounds:
    A. water
    B. methylamine
    C. 1-propyne
    D. 2-pentyne
The correct order of acidity (most acidic first) is:
(a) C,D,B,A
(b) A,B,C,D
(c) B,C,A,D
(d) A,C,B,D
(e) none of the above is correct

Second Examination (E)   Answer Set

1.(b) 2.(d) 3.(c) 4.(d) 5.(a) 6.(c) 7.(b) 8.(c) 9.(b) 10.(c) 11.(d) 12.(b)
13.(e) 14.(d) 15.(a) 16.(d) 17.(c) 18.(d) 19.(d) 20.(d)

46

Third Examination (A)   One Hour

1. (25 pts, 15 min) Alkyl halides may undergo elimination or substitution reactions according to the E1, E2, S_N1, or S_N2 mechanisms. Consider the reaction of an anion, X^-, with an alkyl iodide, RCH_2I, in the presence of 80% methanol, 20% water. Explain the effect of the following changes on the rates of the E1 and S_N2 processes.
   (a) doubling the concentration of X^-
   (b) changing from R = CH_3 to R = $(CH_3)_3C$
   (c) changing from RCH_2I to RCH_2Cl
   Also explain the effects of the following changes on competing E_2 and S_N2 pathways:
   (d) doubling the concentration of X^-
   (e) changing from R = propyl to R = isopropyl
   Finally, explain the effect on competing S_N2 and S_N1 pathways of changing the solvent to 20% methanol, 80% water.

2. (25 pts, 15 min) All parts of this question relate to nmr spectroscopy.
   (a) A compound $C_8H_{18}O$ contains a single sharp peak in its nmr spectrum. What is the structure of the compound?
   (b) A compound $C_5H_{11}Br$ gives the following nmr spectrum. What is its structure?

δ value	no. of protons	splitting
0.80	6	doublet
1.02	3	doublet
2.05	1	multiplet
3.53	1	multiplet

   (c) Deduce the structure of $C_3H_6Cl_2$ from its nmr spectrum.

δ value	no. of protons	splitting
2.20	2	quintet
3.80	4	triplet

   (d) Deduce the structure of C_3H_8O from its nmr spectrum.

δ value	no. of protons	splitting
1.20	6	doublet
1.60	1	broad singlet
4.00	1	septet

3. (25 pts, 15 min) Give the major products of the following reactions.
   (a) $(CH_3)_2C=CHCH_3 + B_2H_6$, then H_2O_2 and $OH^- \longrightarrow$

   (b) $(CH_3)_3CCH(OH)CH_3 + H_2SO_4$, heat $\longrightarrow$

   (c) $H_3C-\langle\bigcirc\rangle-MgBr + CH_3\overset{O}{\overset{\|}{C}}-CHCH_3 \xrightarrow{\text{acid work up}}$

   (d) $H_3C-\langle\bigcirc\rangle-SO_2Cl + C_2H_5OH \xrightarrow{\text{aq. } OH^-}$

4. (25 pts, 15 min) Show how each of the following conversions may be accomplished. More than one step may be required for each.

(a)

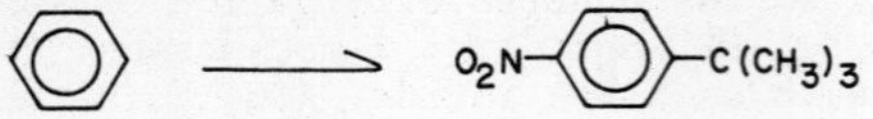
O2N
C(CH3)3

(b)

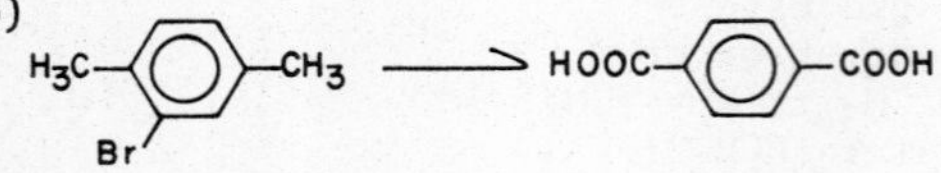
H3C
CH3
Br
HOOC
COOH

(c)

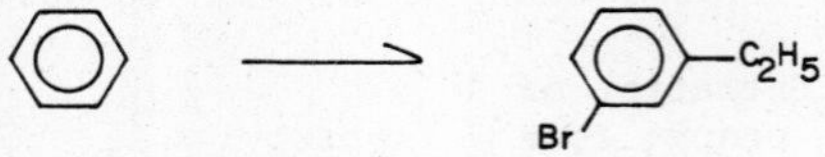
C2H5
Br

(d)

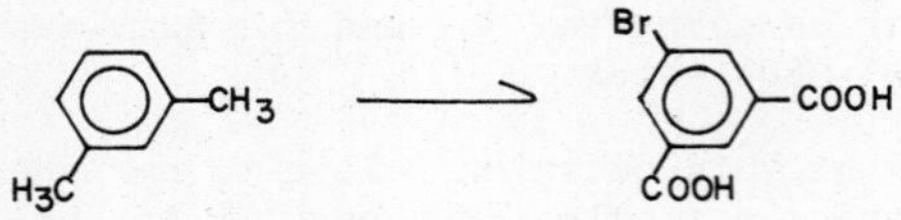
CH3
H3C
Br
COOH
COOH

Third Examination (A)   Answer Set

1. First part

Change	Effect on E1 process	Effect on S_N2 process
(a)	No effect; rate depends only on [halide ion]	Rate will double
(b)	Rate will increase; more stable carbonium ion is produced and a more stable alkene after loss of a proton.	Rate will decrease owing to steric hindrance.
(c)	Rate will decrease; I^- is a better leaving group than Cl^-.	Rate will decrease; I^- is a better leaving group than Cl^-.

Second part
(d) Doubling the concentration of methoxide ion would increase both the rate of elimination and substitution by a factor of two. The relative amounts of substitution and elimination products would therefore remain the same.
(e) Changing from R = n-propyl to R = isopropyl would tend to favor elimination over substitution owing to increased steric hindrance presented to a potential nucleophile.

Third part
The increase in the relative amount of water to methanol makes the solvent more polar. This will favor the formation of a carbonium ion intermediate; thus the rate of the S_N2 reaction will be decreased and the rate of the S_N1 reaction increased.

2. (a) Since the compound contains just one peak in its nmr spectrum, there must be no observable spin-spin splitting, and all protons must be equivalent. The formula $C_8H_{18}O$ therefore suggests two t-butyl groups and an oxygen atom. The required structure therefore is

$$(CH_3)_3C-O-C(CH_3)_3$$

(b) The 6H and 3H doublets at δ0.80 and δ1.02 suggest the following structural units:

6H doublet — $(CH_3)_2CH-$ and $CH_3C\overset{\frown}{H}$

Thus all five carbons are accounted for. The bromine clearly must be attached to the $CH_3CH=$ unit, so the overall structure must be

$$(CH_3)_2CHCH(Br)CH_3$$

(c) The 2H quintet at δ2.20 shows two protons spin-coupled to four equivalent neighboring protons, suggesting the structural unit

 2H quintet ⸻⟶ $H_2C(CH_2X)_2$

Clearly then, if one makes X = Cl, one has the required structure.

$$H_2C(CH_2Cl)_2$$

The 4H triplet then corresponds to the four -CH$_2$Cl protons spin-coupled to the two H$_2$C= protons.

(d) The nmr spectrum suggests the following:

  6H doublet, δ1.20 (C$\underline{H}_3$)$_2$CH—

  1H singlet, δ1.60 —O$\underline{H}$

  1H septet, δ4.00 (CH$_3$)$_2$C$\underline{H}$—

Thus the required structure is clearly (CH$_3$)$_2$CHOH.

3. (a)   (CH$_3$)$_2$CHCH(OH)CH$_3$

This alcohol is formed <u>via</u> an intermediate organoboron compound. The overall process involves addition of H$_2$O to the double bond with anti-Markownikov orientation.

(b) $(CH_3)_3CCH(OH)CH_3 \xrightarrow{acid} (CH_3)_3C\overset{+OH_2}{C}HCH_3 \xrightarrow{-H_2O} (CH_3)_3\overset{+}{C}CHCH_3$

                         ↘ −H$^+$

             $(CH_3)_2C=C(CH_3)_2$ ⟵ $(CH_3)_2\overset{+}{C}CH(CH_3)_2$

(c) H$_3$C–⟨C$_6$H$_4$⟩–MgBr CH$_3$CH–CHCH$_3$ (epoxide) $\xrightarrow{\text{acid work up}}$ H$_3$C–⟨C$_6$H$_4$⟩–CH(CH$_3$)CH(OH)CH$_3$

(d) H$_3$C–⟨C$_6$H$_4$⟩–SO$_2$Cl $\xrightarrow{C_2H_5OH}$ H$_3$C–⟨C$_6$H$_4$⟩–SO$_2$OC$_2$H$_5$

4. (a) ⟨C$_6$H$_6$⟩ $\xrightarrow[AlCl_3]{(CH_3)_3CCl}$ ⟨C$_6$H$_5$⟩–C(CH$_3$)$_3$ $\xrightarrow[H_2SO_4]{HNO_3}$ O$_2$N–⟨C$_6$H$_4$⟩–C(CH$_3$)$_3$

(b) H$_3$C–⟨C$_6$H$_3$(Br)⟩–CH$_3$ $\xrightarrow[Et_2O]{Mg}$ H$_3$C–⟨C$_6$H$_3$(MgBr)⟩–CH$_3$ $\xrightarrow{H_2O}$ H$_3$C–⟨C$_6$H$_4$⟩–CH$_3$ $\xrightarrow{aq.KMnO_4}$ HOOC–⟨C$_6$H$_4$⟩–COOH

(c) Br-C₆H₄-CH₂CH₃ →[Zn(Hg), HCl]← Br-C₆H₄-COCH₃ →[Br₂, Fe]← C₆H₅-COCH₃

C₆H₆ →[CH₃COCl, AlCl₃]

(d) m-xylene →[aq. KMnO₄] m-C₆H₄(COOH)₂ →[Br₂, Fe] 5-bromo-1,3-benzenedicarboxylic acid

Again the correct sequence of operations is essential. The last step would be difficult because there are two deactivating carboxylic acid groups on the ring.

Third Examination (B)   One Hour

1. (25 pts, 15 min) Show how the following transformations could be effected. In addition to the given starting materials one may use any inorganic compounds and any alkyl halides containing three or less carbon atoms.

   (a) cyclohexane (benzene ring) → CH$_2$=CH–CH$_2$–CH$_2$–CH=CH$_2$ (hexadiene)

   (b) benzene → 4-HOOC, 3-NO$_2$, 1-Br substituted benzene (HOOC, NO$_2$, Br on ring)

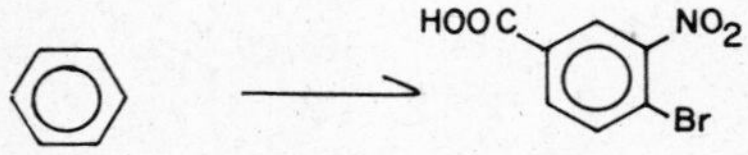

   (c) H_2C_2 and C_2H_5OH → $(C_2H_5)_2C(OH)CH_2CH_2CH_3$

   (d) benzene → 1,2-dinitro-3-chlorobenzene (NO$_2$, Cl, NO$_2$ on ring)

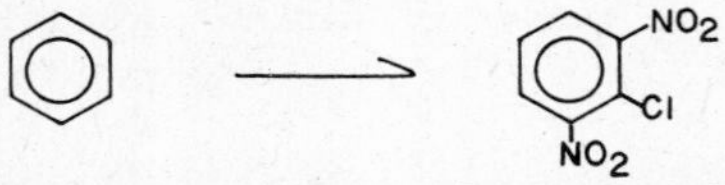

2. (25 pts, 15 min) Identify compound **A** from the following data:
   Empirical formula: C_3H_6O
   nmr Spectrum:

δ value	relative area	splitting
1.2	6	singlet
2.2	3	singlet
2.6	2	singlet
4.0	1	singlet

   ir Spectrum:
   signals at 1710 cm^{-1} and 3400 cm^{-1} (partial)

   Why do acidic protons normally absorb "downfield" in nmr spectroscopy? Bearing this in mind, offer an explanation for the fact that acetylenic protons, which are, of course, acidic, absorb at a <u>relatively</u> high field position (between δ = 2 and δ = 3).

3. (16 pts, 10 min) Benzene reacts with an ionic species X$^+$ as follows:
   $$C_6H_6 + X^+ \longrightarrow C_6H_5X + H^+$$
   The overall reaction is exothermic. The mechanism for this reaction is as follows:

   Step 1 $C_6H_6 + X^+ \underset{k_2}{\overset{k_1}{\rightleftarrows}} [C_6H_5\overset{H}{\underset{X}{\diagup\diagdown}}]^+$

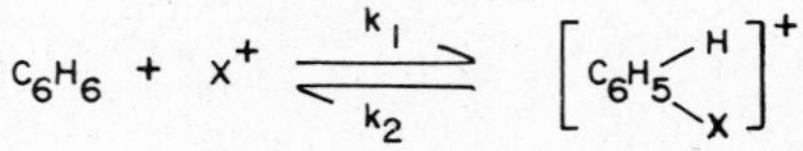

   Step 2 $[C_6H_5\overset{H}{\underset{X}{\diagup\diagdown}}]^+ \underset{k_4}{\overset{k_3}{\rightleftarrows}} C_6H_5X + H^+$

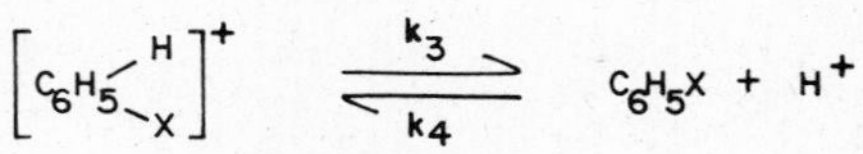

   where k_1, k_2, k_3, and k_4 are the rate constants for the different forward and back reactions. The back reaction for step 2 is negligibly small, and the following order has been deduced:
   $k_3 > k_2 > k_4 > k_1$

Draw a reaction co-ordinate diagram, labelling all intermediates, reactants, final products, etc. Which step is rate determining? Explain.

4. (17 pts, 10 min) Supply structures for the lettered compounds.

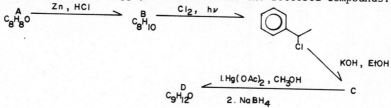

5. (17 pts, 10 min) Complete each of the following reactions. If no reaction occurs, so state.

(a)

H_3C—⟨⟩—⟨⟩—NO_2 $\xrightarrow{CH_3COCl, AlCl_3}$ A

(b)

[PhC(O)OPh] $\xrightarrow{HNO_3, H_2SO_4}$ B

(c)

O_2N—⟨⟩—⟨⟩—CN $\xrightarrow{CH_3Cl, AlCl_3}$ C

(d)

⟨⟩ $\xrightarrow{fuming\ H_2SO_4}$ D $\xrightarrow{excess\ Br_2, Fe}$ E $\xrightarrow{acid, steam}$ F

Third Examination (B)　　　　　　　　　　　　　　　　　　　　　Answer Set

1. (a) 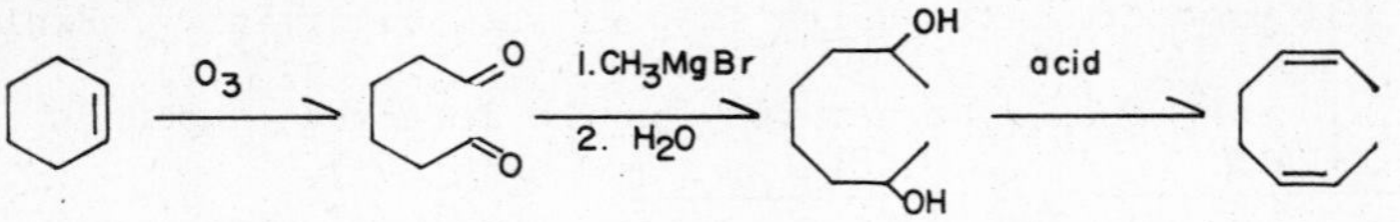 cyclohexene $\xrightarrow{O_3}$ dialdehyde $\xrightarrow[\text{2. H}_2\text{O}]{\text{1. CH}_3\text{MgBr}}$ diol $\xrightarrow{\text{acid}}$ cyclooctadiene

   (b) 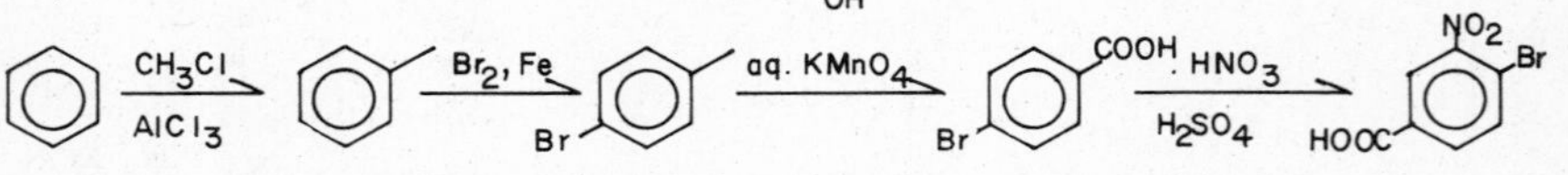 benzene $\xrightarrow[\text{AlCl}_3]{\text{CH}_3\text{Cl}}$ toluene $\xrightarrow{\text{Br}_2,\text{Fe}}$ p-bromotoluene $\xrightarrow{\text{aq. KMnO}_4}$ p-bromobenzoic acid $\xrightarrow[\text{H}_2\text{SO}_4]{\text{HNO}_3}$ product

   (c) 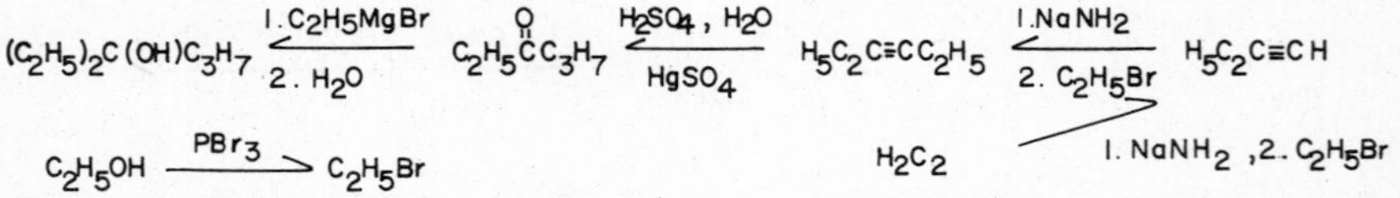 $(C_2H_5)_2C(OH)C_3H_7 \xleftarrow[\text{2. H}_2\text{O}]{\text{1. C}_2\text{H}_5\text{MgBr}} C_2H_5\overset{O}{\overset{\|}{C}}C_3H_7 \xleftarrow[\text{HgSO}_4]{\text{H}_2\text{SO}_4,\text{H}_2\text{O}} H_5C_2C\equiv CC_2H_5 \xleftarrow[\text{2. C}_2\text{H}_5\text{Br}]{\text{1. NaNH}_2} H_5C_2C\equiv CH$

   $C_2H_5OH \xrightarrow{PBr_3} C_2H_5Br$　　　　　　　　　　　　$H_2C_2 \xrightarrow{\text{1. NaNH}_2, \text{2. C}_2\text{H}_5\text{Br}}$

   (d) 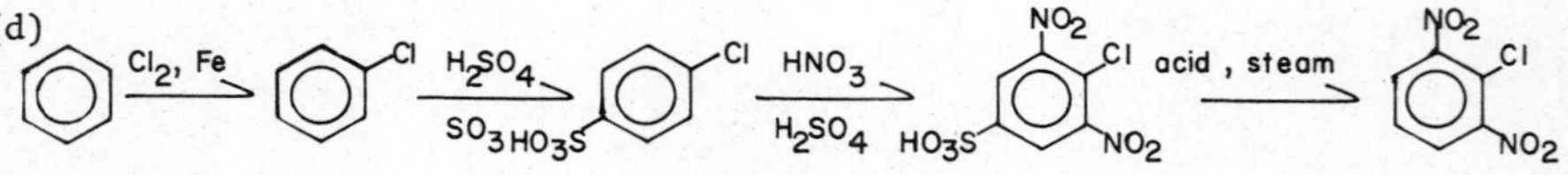 benzene $\xrightarrow{\text{Cl}_2,\text{Fe}}$ chlorobenzene $\xrightarrow[\text{SO}_3\text{H}]{\text{H}_2\text{SO}_4}$ $\xrightarrow[\text{H}_2\text{SO}_4 \; \text{HO}_3\text{S}]{\text{HNO}_3}$ $\xrightarrow{\text{acid, steam}}$ product

   This last route illustrates the use of the -SO₃H group as a blocking group. It also serves the purpose of directing further substitution into a position <u>meta</u> to itself. Finally, it is worth noting that one is relying on the reversible nature of the sulfonation reaction in this sequence: its introduction using concentrated sulfuric acid and its removal by heating with aqueous acid.

2. Since the nmr spectrum shows 12 protons, the molecular formula must be "twice" the empirical formula, i.e. $C_6H_{12}O_2$. The ir spectrum shows the presence of -OH and carbonyl functionalities. The following structure fills all the requirements:

   $$HOC(CH_3)_2CH_2COOH$$

   If a proton absorbs "downfield" it means that a smaller magnetic field is required to bring it into resonance compared to other protons. In turn, this implies that the proton is shielded only to a relatively small extent. Shielding, one should recall, is mainly the result of the circulation of electrons around a proton itself, generating a magnetic field, which, at the proton, opposes the applied field. In general, one might imagine that the smaller the electron density in the vicinity of the particular proton, the smaller would be the induced field, and thus the smaller the shielding. Such protons should therefore absorb downfield in the absence of any other factors.

   Consider a typical acidic proton, e.g. one attached to a carboxylic acid:

   $$RCOOH$$

   The proton is attached to an electronegative atom, one that draws electron density to itself. This would produce just the sort of weak shielding effect seen above. In general, an acidic proton is one around which there is diminished electron density. This generally causes a "downfield" resonance. To explain the apparently anomolous chemical shifts of acetylenic protons one must look beyond the electron density about the proton itself. Since the acetylenic protons absorb in the nmr spectrum at a much higher field than might be expected for acidic protons, it must be true that these protons experience a substantial secondary induced field in a direction opposite to the applied field. This would mean that a greater applied field than otherwise expected would be required to bring these protons into resonance. This secondary field is caused by circulation of electrons in the C-C triple bond as shown below.

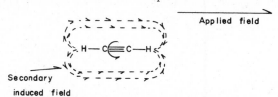

3. Step 1 must be the slow step since $k_2 > k_1$. Thus E_A, the energy of activation for step 1 must be larger than E_A for step 2. One can use this information, along with the given fact that the reaction is exothermic, to construct the reaction coordinate diagram.

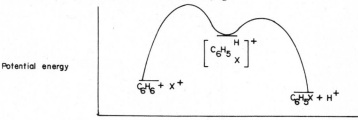

4. This rather straight-forward problem is solved readily by working from the single specified structure, 1-chloro-1-phenylethane. One sees that it is formed from a free-radical chlorination reaction, so compound B must be ethylbenzene, the chlorination occurring at the benzylic position via the resonance stabilized radical, $C_6H_5\dot{C}HCH_3$.

Compound B is formed from A by a Zn/HCl reduction. Thus one can interpret this reaction as a Clemmensen reduction of acetophenone.

Turning now to the structures of C and D, one can expect that 1-chloro-1-phenylethane will undergo dehydrohalogenation upon treatment with KOH/ethanol to give styrene.

[Reaction: 1-chloroethylbenzene → styrene]

Methoxymercuration-demercuration of C will have the overall effect of adding methanol across the double bond with Markownikov orientation.

Styrene $\xrightarrow{\text{1. Hg(O}_2\text{CCF}_3)_2, \text{ CH}_3\text{OH}}_{\text{2. NaBH}_4}$ PhCH(OCH$_3$)CH$_3$

5. (a) A = 3-acetyl-4'-nitrobiphenyl (H$_3$CCO substituent)

   (b) B = 4-nitrophenyl benzoate

   (c) C = no reaction

   (d) D = p-toluenesulfonic acid; E = 3,5-dibromo-4-methylbenzenesulfonic acid; F = 1,3-dibromo-2-methylbenzene

Third Examination (C)   One Hour

1. (25 pts, 15 min) Compare the rates of the competing reactions in each of the five parts of this problem. Offer brief but concise explanations which should include stereochemical drawings where appropriate.
   (a) Elimination of HBr from the cis- and trans-isomers of 4-isopropylcyclohexylbromide.
   (b) $(CH_3)_2CHCl$ — KI, EtOH →
        $(CH_3)_2CHOH$ — KI, EtOH → $(CH_3)_2CHI$
   (c) [structure: isopropylbenzene with Br] — KOH, EtOH → [isopropenylbenzene] / KOH, EtOH → [HO-substituted product]
   (d) $CH_3CH_2CH(Br)CH_3$ + NaOEt → [alkene products]
   (e) S_N1 displacement of bromide ion from [allyl bromide, norbornyl bromide, benzyl bromide, vinyl-type bromide]

2. (25 pts, 15 min) Supply the missing structures and/or reagents.
   (a) [cyclohexene] $\xrightarrow{Hg(OAc)_2, H_2O}$ A ($C_9H_{16}O_2Hg$) $\xrightarrow{B}$ C ($C_7H_{14}O$)
   (b) [benzene] $\xrightarrow{D \text{ and } E}$ [ethylbenzene]
   (c) [phenyl benzyl ether, Ph-O-CH$_2$-Ph] $\xrightarrow{HNO_3, H_2SO_4}$ F (major mononitration product)
   (d) [phenyl benzyl ketone, Ph-CO-CH$_2$-Ph] $\xrightarrow{H_2SO_4, SO_3}$ G (major product)
   (e) $FCH_2CH_2CH_2Br$ $\xrightarrow[CH_3OH]{\text{excess NaOCH}_3}$ H

3. (25 pts, 15 min) Deduce the structures of the following compounds from their proton nmr spectra.

   Compound I: C_3H_6O quintet at $\delta = 2.7$, triplet at $\delta = 4.7$.

   Compound II: C_4H_8O triplet(3H) at $\delta = 1.1$, singlet(3H) at $\delta = 2.1$, quartet at $\delta = 2.5$.

57

Compound III: doublet(6H) at δ = 1.0, multiplet(1H) at δ = 2.0, doublet(2H) at δ = 3.3.

4. (25 pts, 15 min) Show how one could synthesize the following compounds, starting from benzene, and using any required inorganic substances, solvents, and other organic compounds containing three or fewer carbons.

(a) 4-Cl-C$_6$H$_4$-C(OH)(H)-CH$_2$OH

(b) 2,4-dinitrobenzoic acid (COOH with NO$_2$ ortho and O$_2$N para)

(c) C$_6$H$_5$-CH$_2$-C≡C-CH$_3$

Third Examination (C) Answer Set

1. (a) The most stable chair conformations of the two compounds are:

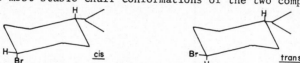

In the <u>cis</u> compound, the bromine atom occupies an axial position. Thus it bears the correct <u>anti</u> relationship to one of the hydrogen atoms on adjacent carbon atoms for efficient E2 elimination. The bromine in the <u>trans</u> compound can get into an axial position only if the ring flips, but this would place the bulky isopropyl group in an axial position, introducing unfavorable 1,3-diaxial interactions between the isopropyl group and ring hydrogens. Thus, one can expect the <u>trans</u> isomer to eliminate HBr more slowly than the <u>cis</u> isomer.

(b) This problem boils down to a comparison of -Cl and OH as leaving groups. Of the two, -OH is by far the poorer leaving group, a fact that stems from the strongly basic nature of the hydroxyl anion. In fact, OH$^-$ is so much more basic than I$^-$ (the attacking species) that the displacement of -OH by -I is not feasible.

(c) The halide is tertiary, indicating that the S_N2 or E2 routes are unfavorable. One must therefore look at a reaction involving a carbonium ion.

3° and benzylic carbonium ion

This carbonium ion can be expected to lose a proton rather easily and thereby form an alkene, especially under conditions as specified: a solvent of low polarity and the use of a strong base (EtOH and KOH). Only if one uses a much weaker base (e.g. water) would one expect the formation of the alcohol to become competitive.

(d) The formation of 2-butene would arise from Saytzeff orientation, while 1-butene would correspond to Hofmann elimination in the dehydrohalogenation. For simple alkyl halides, Hofmann orientation becomes dominant only if the halogen is fluorine. Thus one expects 1-butene to be only a minor product of the reaction. To decide which of the isomers of 2-butene will dominate (i.e. be formed quicker), one should look at the conformations of the starting material which place the eliminated H and Br in an <u>anti</u> relationship.

Conformation II, which would lead to the <u>cis</u> alkene, is seen to have an unfavorable CH_3/CH_3 <u>gauche</u> interaction. Thus one expects elimination to occur chiefly from conformation I, resulting in the <u>trans</u> alkene.

(e) One is required to compare the ease of S_N1 displacement of Br from the following:

<pre>
    Br   Br
   ⫽   ⟋Br   ⟋⟍
    1   2   3   Br 4
</pre>

Displacement of Br from 1 is extremely difficult under any circumstances; the C-halogen bond in vinyl halides is simply too strong to be broken easily (why?). Compound 2 is a <u>tertiary</u> halide, so one might imagine that S_N1 displacemnt of Br would be relatively easy. However, this in fact is not the case, because the resulting carbonium ion could not achieve planarity owing to the bicyclic ring system imposing a rigidity on the molecule. Thus the carbonium ion would be extremely unstable and difficult to form. Compound 3 is a primary halide, so it is more likely to react by an S_N2 than an S_N1 mechanism. Only compound 4, an allylic bromide, would be expected to form a carbonium ion relatively easily, since a resonance stabilized species results.

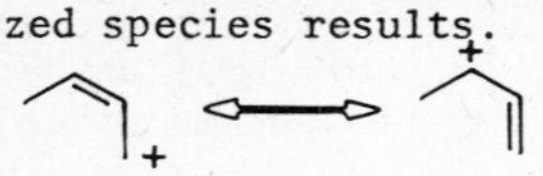

$$\diagdown\!\diagup^+ \longleftrightarrow \diagup\!\diagdown^+$$

2. (a)

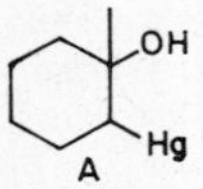

   A: cyclohexane with OH and Hg substituents   B: NaBH$_4$ C: cyclohexanol

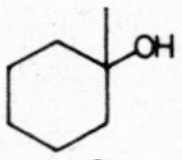

   (b) D: C_2H_5Cl E: $AlCl_3$

   (c) The phenyl ring attached to the oxygen atom is more highly activated than the one attached to $-CH_2O-$. Thus the major product will be:

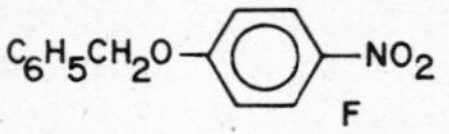

   $C_6H_5CH_2O-\!\!\bigcirc\!\!-NO_2$
   F

   (d) The phenyl ring attached to the carbonyl group is deactivated toward electrophilic substitution. Therefore the other phenyl ring will be substituted preferentially, mainly at the <u>para</u> position.

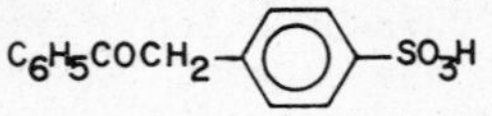

   $C_6H_5COCH_2-\!\!\bigcirc\!\!-SO_3H$
   G

   (e) Fluorine will not be displaced as it is too poor a leaving group. Fluoride ion is a very strong base, a result of the weak acidity of the conjugate acid, HF. Thus the expected reaction is:

   $FCH_2CH_2CH_2OCH_3$
   H

3. Compound I: The quintet suggests the unit:

   $-CH(CH_2-)_2$ or $H_2C(CH_2-)_2$

   Since one also sees a triplet and nothing else in the spectrum, one must choose B. This part structure contains three carbon atoms and six hydrogens, so there is only one oxygen to place. Thus the structure of I is:

[structure: oxetane ring with O]

Compound II:

[structure: methyl ethyl ketone / propan-2-one derivative with C=O]

Compound III:

[structure: isobutyl bromide (CH₃)₂CHCH₂Br]

The two methyl groups appear as a doublet owing to coupling with the single adjacent -CH- proton. Likewise the -CH₂Br protons appear as a doublet because of coupling to the same -CH- proton. The -CH- proton will be split first of all into a septet by coupling with the six -CH₃ protons, then each member of the septet is split into a triplet by coupling to the -CH₂Br protons. Thus a multiplet is seen for the -CH- proton.

4. (a) benzene + (CH₃)₂CHCl / AlCl₃ → isopropylbenzene → Cl₂, Fe → o-chloroisopropylbenzene → Cl₂, heat → p-chloro compound with CCl →

then → Cl–C₆H₄–C(OH)(CH₃)₂ (HO, OH) ← H₂O₂, HCOOH ← Cl–C₆H₄–C(=CH₂)CH₃ ← KOH, EtOH

(b) benzene → CH₃Cl, AlCl₃ → toluene → HNO₃ / H₂SO₄ → 2,4-dinitrotoluene (O₂N, NO₂) → hot aq. K₂Cr₂O₇ → 2,4-dinitrobenzoic acid (COOH, O₂N, NO₂)

(c) benzene → CH₃Cl, AlCl₃ → toluene → Br₂, heat → benzyl bromide (CH₂Br) → NaCCCH₃ → PhCH₂C≡CH

Third Examination (D)   One Hour

1. (25 pts, 15 min) (a) Give the structure of the product expected from the reaction of benzene with propylene in the presence of a strong acid. Show a mechanism.
   (b) Give a reaction-progress vs. potential energy graph for the reaction of NO_2^+ with benzene. Label the intermediate and provide a structure for it.
   (c) Consider the reaction of anisole with bromine, resulting in <u>para</u> substitution. Show all reasonable resonance structures for the initially formed intermediate.

2. (25 pts, 15 min) Consider the reaction:

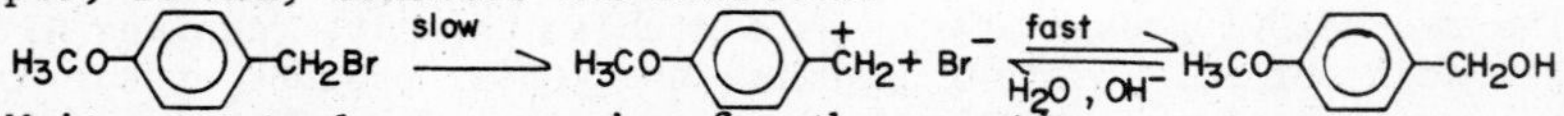

   $H_3CO-C_6H_4-CH_2Br \xrightarrow{slow} H_3CO-C_6H_4-CH_2^+ + Br^- \xrightleftharpoons[H_2O, OH^-]{fast} H_3CO-C_6H_4-CH_2OH$

   (a) Write a rate law expression for the reaction.
   (b) What would be the effect on the <u>rate constant</u> of doubling the concentration of OH^-?...of p-methoxybenzyl bromide?
   (c) Repeat part (b), substituting <u>rate</u> for <u>rate constant</u>.
   (d) Draw a reaction progress vs. potential energy diagram for the process.
   (e) Would the reaction proceed more quickly or more slowly if benzyl bromide were used in place of p-methoxybenzyl bromide? Explain.

3. (25 pts, 15 min) Deduce the structures of compounds A-E from the data.
   A $C_9H_{10}O$ ir absorption at ca. 1715 cm^{-1}; no absorption above 3100.
                    nmr: 5H singlet at 7.2δ, 2H singlet at 3.5δ, 3H singlet at 1.9δ.
   B $C_{10}H_{14}$ nmr: 5H singlet at 7.2δ, 1H broad at 2.8δ, 2H multiplet at 1.8δ, 3H doublet at 1.3δ, 3H triplet at 1.0δ.
   C $C_{10}H_{14}$ nmr: 5H singlet at 7.2δ, 2H triplet at 2.6δ, 4H multiplet at 1.8δ, 3H triplet at 0.9δ.
   D $C_6H_{14}O$ nmr: 1H singlet at 4.40δ, 1H quartet at 3.40δ, 3H doublet at 1.10δ, 9H singlet at 0.90δ.
                    D is converted to E by the following sequence:

   $D \xrightarrow{acid} C_6H_{12} \xrightarrow[\text{2. Zn, H}_2\text{O}]{\text{1. O}_3} E (C_3H_6O)$

   E C_3H_6O nmr: single peak at 2.1δ.

4. (25 pts, 15 min) Provide structures for the missing compounds or reagents. If a given set of reagents results in no reaction, so indicate.

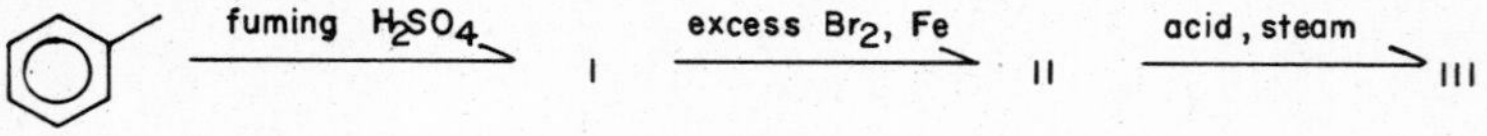

   benzene $\xrightarrow{\text{fuming H}_2\text{SO}_4}$ I $\xrightarrow{\text{excess Br}_2, \text{Fe}}$ II $\xrightarrow{\text{acid, steam}}$ III

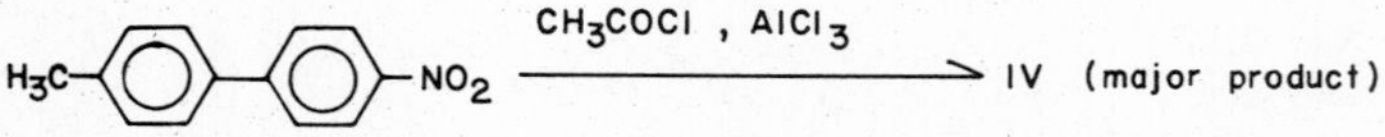

   $H_3C-C_6H_4-C_6H_4-NO_2 \xrightarrow{CH_3COCl, AlCl_3}$ IV (major product)

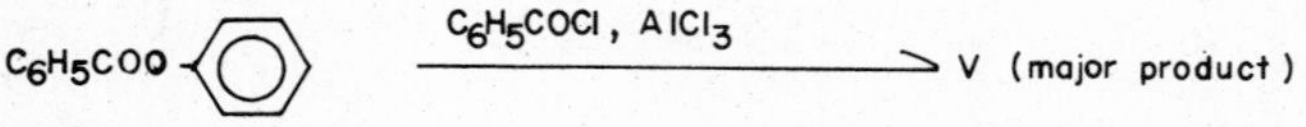

   $C_6H_5COO-C_6H_5 \xrightarrow{C_6H_5COCl, AlCl_3}$ V (major product)

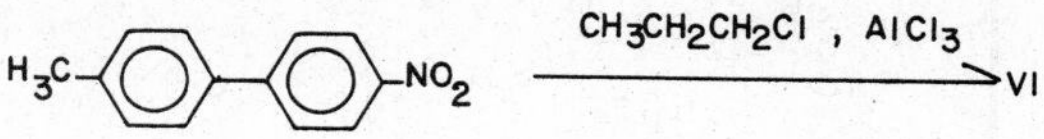

H_3C—⌬—⌬—NO_2 $\xrightarrow{CH_3CH_2CH_2Cl \; , \; AlCl_3}$ VI

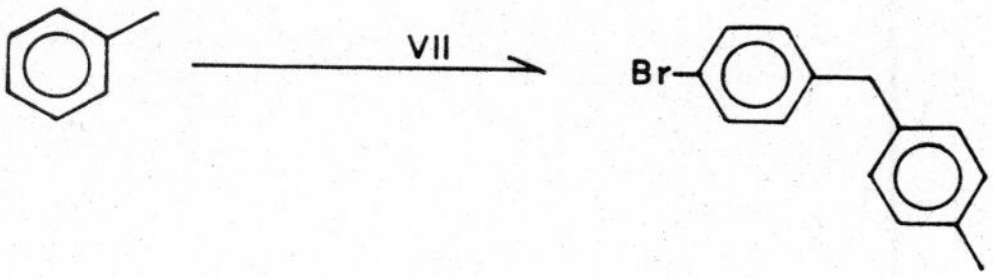

⌬—CH$_3$ $\xrightarrow{VII}$ Br—⌬—CH$_2$—⌬—CH$_3$

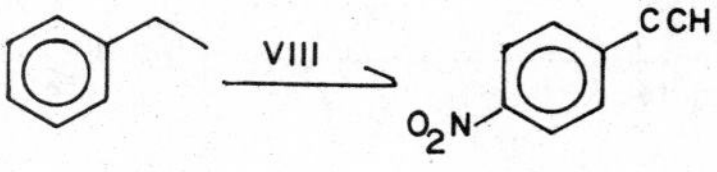

⌬—CH$_2$—⌬ $\xrightarrow{VIII}$ O_2N—⌬—C≡CH

Third Examination (D) Answer Set

1. (a) CH₂=CHCH₃ (propene) + H⁺ → (CH₃)₂CH⁺ ; + C₆H₆ → [cyclohexadienyl cation with isopropyl and H] ; –H⁺ → isopropylbenzene

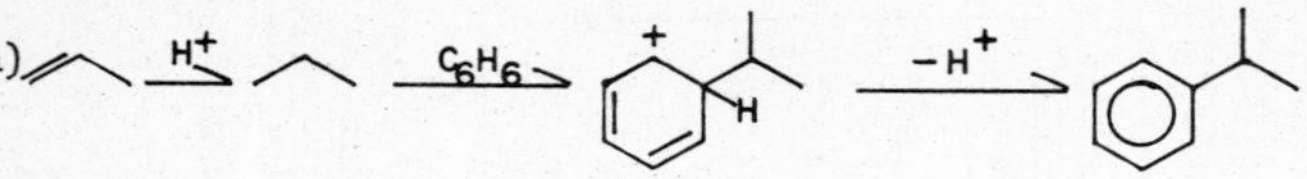

   (b) Potential energy vs. Reaction coordinate diagram showing two-hump profile; arrow points to intermediate corresponding to the σ-complex (cyclohexadienyl cation bearing H and NO₂).

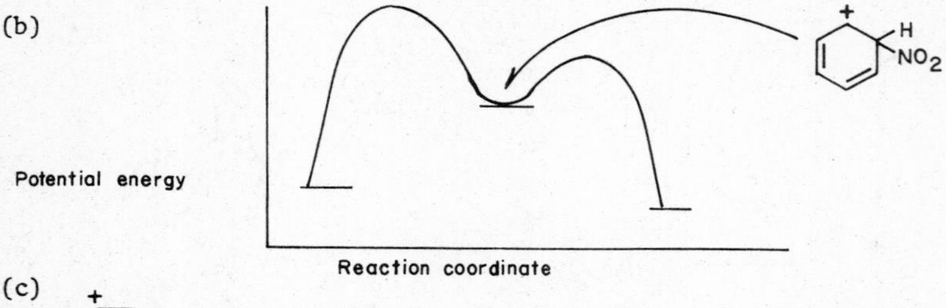

   (c) Resonance structures of the σ-complex for electrophilic bromination of anisole (para attack): four contributing structures showing positive charge delocalized onto ring carbons and onto the oxygen of –OCH₃, with Br and H on the sp³ carbon.

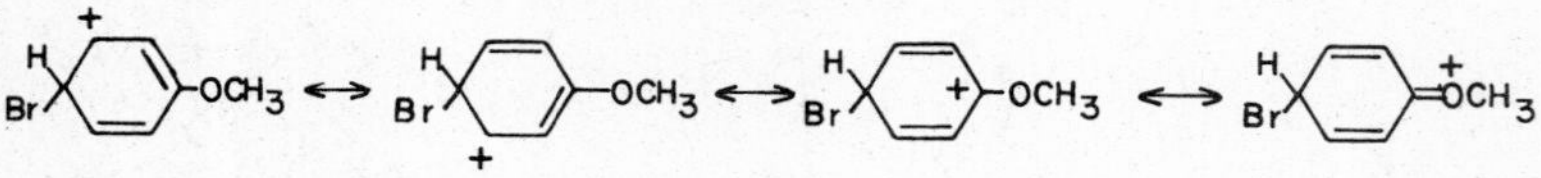

2. (a) Rate = k[p-methoxybenzyl bromide]
   (b) No effect in either case. The rate constant is not dependent on the concentrations.
   (c) The rate would be unaffected by changing [OH⁻], but would double if [p-methoxybenzyl bromide] were doubled.
   (d) Potential energy vs. Reaction coordinate diagram; arrow points to the carbocation intermediate H₃CO–C₆H₄–CH₂⁺.
   (e) The reaction proceeds more quickly with the –OCH₃ substituent present. It helps to stabilize the carbonium ion intermediate through resonance.

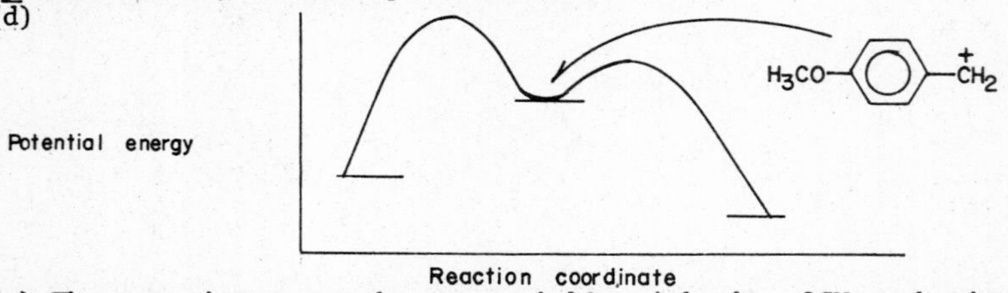

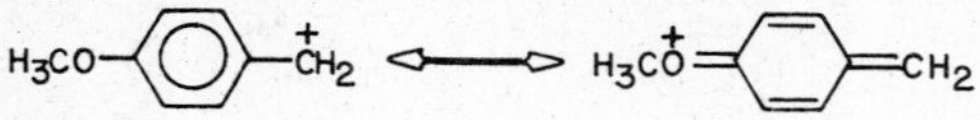

   H₃CO–C₆H₄–CH₂⁺ ↔ H₃C=O⁺=C₆H₄=CH₂

3.
   C₆H₅CH₂COCH₃   C₆H₅CH(CH₃)C₂H₅   C₆H₅CH₂CH₂CH₂CH₃   (CH₃)₃CCH(OH)CH₃   (CH₃)₂C=C(CH₃)₂
       A   B   C   D   E

   Mechanism for D to E conversion:

   $(CH_3)_3CCH(OH)CH_3 \xrightarrow[-H_2O]{H^+} (CH_3)_3CCH^+CH_3 \xrightarrow{\text{methyl shift}} (CH_3)_2C^+CH(CH_3)_2 \xrightarrow{-H^+} (CH_3)_2C=C(CH_3)_2$

64

4.

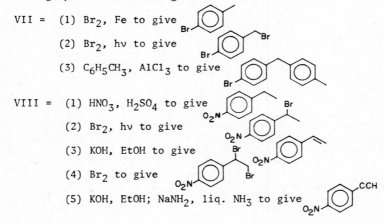

VI = recovered starting materials; Friedel-Crafts acylation will not work on highly deactivated rings.

VII = (1) Br_2, Fe to give
    (2) Br_2, hν to give
    (3) $C_6H_5CH_3$, $AlCl_3$ to give

VIII = (1) HNO_3, H_2SO_4 to give
     (2) Br_2, hν to give
     (3) KOH, EtOH to give
     (4) Br_2 to give
     (5) KOH, EtOH; $NaNH_2$, liq. NH_3 to give

Third Examination (E)   One Hour

1. A compound I exhibits strong infra-red absorption at 1710 cm^{-1} and also shows a very broad absorption band stretching from 2500 to 3500 cm^{-1}. The nmr spectrum of I shows a singlet at 11.7δ (relative area 1), a quartet at 2.2δ, and a triplet at 1.1δ (relative areas 2 and 3 respectively).
   Compound I is:
   (a) 2-butanone
   (b) 5-hydroxybutanoic acid
   (c) acetic acid
   (d) propanoic acid
   (e) p-ethoxybenzoic acid

2. Arrange in order of increasing δ values:
   $CH_3CH_2CH_3$ CH_3CH_3 $(CH_3)_3CH$

        A   B   C
   (a) A,B,C
   (b) C,B,A
   (c) B,A,C
   (d) B,C,A
   (e) C,A,B

3. A compound C_7H_8O can be made to undergo Friedel-Crafts acylation to yield a substance $C_{10}H_{12}O_2$. The nmr spectrum of this latter compound consists of a pair of doublets at 7.0δ and 8.0δ (total of 4H), a singlet at 3.9δ (3H), a quartet at 2.9δ (2H), and a triplet at 1.2δ (3H). The compound $C_{10}H_{12}O_2$ is:
   (a) p-ethoxyacetophenone
   (b) p-methoxypropiophenone
   (c) 1-p-hydroxyphenylbutanone
   (d) 4-(n-propyl)-benzoic acid
   (e) Benzyl propionate

4. A compound $C_9H_{12}O$ exhibits infra-red absorption at 3400 cm^{-1}, but not between 1600 and 2800 cm^{-1}. Its nmr spectrum consists of a singlet at 7.2δ (relative area 5), a triplet at 4.2δ (relative area 1), a singlet at 3.8δ (relative area 1), a quartet at 1.6δ (relative area 2), and a triplet at 0.8δ (relative area 3). The compound is:
   (a) p-hydroxypropyl benzene
   (b) 2-(p-tolyl)-ethanol
   (c) 1-phenylpropanol
   (d) 2-phenylpropanol
   (e) 4-methoxyethyl benzene

5. The two main isotopes of chlorine, ^{35}Cl and ^{37}Cl, have relative abundances of approximately 75% and 25% respectively. Consider now the mass spectrum of p-dichlorobenzene. If the peak at m/e 146 has an intensity of 100 units, what is the expected intensity for the m/e 148 peak?
   (a) 25

(b) 6.3
(c) 50
(d) 66.7
(e) 75

6. Which of the following compounds will absorb uv or visible light of the longest wavelength?
   (a) 1,3-cyclohexadiene
   (b) cyclohexene
   (c) cyclohexane
   (d) 1,4-hexadiene
   (e) bicyclo[2.2.2]octa-2-ene

7. For each pair, pick out the more easily hydrolyzed compound:
   (i) A vinyl chloride or B allyl chloride
   (ii) C ethyl chloride or D ethyl iodide
   (iii) E benzyl bromide or F p-bromotoluene
   (a) A,C,E
   (b) B,D,E
   (c) B,C,F
   (d) A,D,E
   (e) none of the above is correct

8. What is the effect of increasing solvent polarity on S_N1 and S_N2 reactions
   (a) both get faster
   (b) S_N1 gets faster, S_N2 gets slower
   (c) S_N1 gets slower, S_N2 gets faster
   (d) S_N1 gets faster, no effect on S_N2
   (e) none of the above is correct

9. Consider the following alcohols:
   A benzyl alcohol
   B 2-propanol
   C 1-propanol
   D t-butanol
   Which will give a reaction with the Lucas reagent (conc. HCl, $ZnCl_2$)?
   (a) A,D
   (b) A,B,D
   (c) B,D
   (d) C
   (e) A,C

10. Consider an experiment in which (R)-2-iodooctane is treated with radioactive iodide, S_N2 conditions prevailing. Suppose that the rate of racemization is R_1 and the rate of incorporation of radioactive iodide is R_2. Which of the following is correct?
    (a) $R_1/R_2 = 2$
    (b) $R_2/R_1 = 2$
    (c) $R_2/R_1 = 1$
    (d) $R_1/R_2 = 4$

(e) R_2/R_1 is greater than 10^3

11. In each case select the conditions which favor substitution over elimination.
    (i) 1-bromopentane with $\underline{A}$ aq. NaOH or $\underline{B}$ EtOH
    (ii) $\underline{C}$ 1-bromopentane or $\underline{D}$ 3-bromopentane with aq. KOH
    (iii) 2-bromopentane with aq. KOH; $\underline{E}$ warm or $\underline{F}$ hot
    (a) A,C,F
    (b) B,D,E
    (c) A,C,E
    (d) B,C,E
    (e) none of the above is correct

12. Which of the following constitutes a useful synthesis of m-nitrobenzoic acid from benzene?
    (a) benzene + (1) HNO_3, H_2SO_4 (2) $CH_3Cl/AlCl_3$ (3) $KMnO_4$, OH^-, heat, (4) H^+
    (b) benzene + (1) $CH_3Cl/AlCl_3$ (2) HNO_3, H_2SO_4 (3) $KMnO_4$, OH^-, heat, (4) H^+
    (c) benzene + (1) $CH_3Cl/AlCl_3$ (2) $KMnO_4$, OH^-, heat (3) H^+ (4) HNO_3, H_2SO_4
    (d) benzene + (1) Cl_2, Fe (2) HNO_3, H_2SO_4 (3) Mg, Et_2O (4) CO_2 (5) H^+
    (e) benzene + (1) CH_3COCl, $AlCl_3$ (2) HNO_3, H_2SO_4, heat (3) OH^-

13. Which of the hydrogen atoms (i)-(v) in the structure below is most easily abstracted by a bromine atom?
    (a) (i)
    (b) (ii)
    (c) (iii)
    (d) (iv)
    (e) (v)

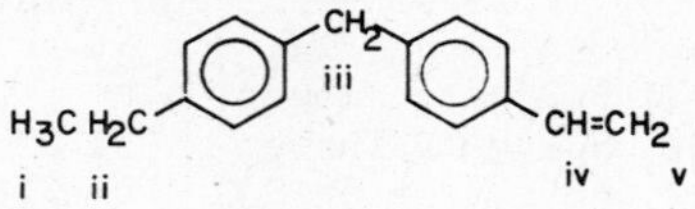

14. Consider the following compounds:
    $\underline{A}$ $C_6H_5NHCOCH_3$
    $\underline{B}$ $C_6H_5COC_2H_5$
    $\underline{C}$ $C_6H_5NH_2$
    $\underline{D}$ C_6H_6
    If one wishes to list these in order of increasing reactivity to NO_2^+, the correct order is:
    (a) A,C,D,B
    (b) D,B,C,A
    (c) D,B,A,C
    (d) B,D,C,A
    (e) none of the above is correct

15. Which of the following alcohols will undergo dehydration most readily?
    (a) 1-phenyl-2-butanol
    (b) 4-phenylbutanol
    (c) 1-phenylbutanol
    (d) 2-phenyl-2-butanol
    (e) 3-phenyl-2-butanol

16. Which of the following compounds is most acidic?

(a) triphenylmethane
(b) toluene
(c) benzene
(d) phenylcyclohexane
(e) cyclopropene

17. In each case pick the compound with the higher boiling point.
    (i) A p-ethyltoluene or B ethoxybenzene
    (ii) C ethyl acetate or D butanoic acid
    (iii) E butanoic acid or F 1-butanol
(a) A,C,E
(b) A,D,E
(c) B,D,F
(d) A,D,E
(e) none of the above is correct

18. When 1,2-dimethylcyclopentene is subjected to hydroboration-oxidation with (1) $(BH_3)_2$ (2) H_2O_2, OH^-, the observed product is:
(a) trans-1,2-dimethylcyclopentanol
(b) cis-1,2-dimethylcyclopentanol
(c) 2,2-dimethylcyclopentanol
(d) cis-2-methylcyclohexanol
(e) trans-2-methylcyclohexanol

19. Give the product of the following reaction sequence.
    toluene + (1) acetyl chloride/$AlCl_3$ (2) Zn(Hg), HCl
(a) 4-ethyltoluene
(b) 3-ethyltoluene
(c) 3-methylbenzyl alcohol
(d) 4-methylbenzyl alcohol
(e) acetophenone

20. Give the product of the following sequence.
    1-phenylpropane + (1) Br_2, heat (2) KOH, EtOH (3) Br_2, CCl_4
    (4) 2 moles of $NaNH_2$ (5) H_2O, $HgSO_4$, H_2SO_4
(a) 1-(p-hydroxyphenyl)butanone
(b) phenyl ethyl ketone
(c) benzyl methyl ketone
(d) 1-phenyl-2-ethylcyclopropane
(e) 1-phenyl-2-hydroxypropane

Third Examination (E)   Answer Set

1.(c) 2.(c) 3.(a) 4.(c) 5.(d) 6.(a) 7.(b) 8.(b) 9.(b) 10.(a) 11.(c) 12.(c)

13.(c) 14.(e) 15.(d) 16.(a) 17.(b) 18.(b) 19.(a) 20.(b)

Final Examination (A)   Two and One-Half Hours

1. (50 pts, 25 min) Give the major organic products of each of the following reactions. Specify isomers where appropriate. If a reaction gives a significant yield of more than one product it must be stated and all structures given. If no reaction is expected it should be so stated. Assume usual work-up conditions in all cases.

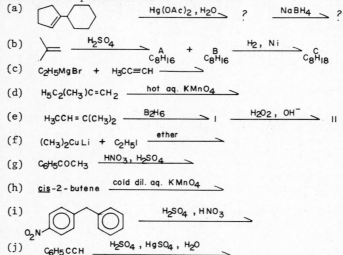

2. (20 pts, 10 min) Compound $\underline{A}$ contains 60.0% C and 13.3% H and has a molecular weight of 60. $\underline{A}$ reacts with oxidizing agents to give an aldehyde and then, on further oxidation, a carboxylic acid. Treatment of $\underline{A}$ with potassium bromide and sulfuric acid gives $\underline{B}$ which reacts with KOH/EtOH to give a hydrocarbon $\underline{C}$. $\underline{C}$ reacts with HBr to give $\underline{D}$ which contains 65.0% Br and which, on hydrolysis, gives $\underline{E}$, a compound which is isomeric with $\underline{A}$. What are the structures of compounds $\underline{A}$-$\underline{E}$? At. Wts.: H = 1, C = 12, Br = 80

3. (15 pts, 8 min) Under catalytic conditions, a hydrocarbon, $C_{10}H_{16}$, absorbs one mole of hydrogen. It has present niether methyl, ethyl, or other alkyl substituent group and on ozonolysis gives a symmetrical diketone of the formula $C_{10}H_{16}O_2$. What are the structures of the hydrocarbon and diketone?

4. (32 pts, 13 min) Write structures fitting each part of the following:
(a) A Fischer projection corresponding to the following Newman projection:

Also indicate R/S designations at each chiral carbon.

(b) An optically active compound, $C_8H_{10}O$, which has a broad band in its ir spectrum centered at 3300 cm^{-1} and the following nmr spectrum: broad singlet at 7.18δ (5H), quartet at 4.65δ (1H), singlet at 2.76δ (1H), and doublet at 1.32δ (3H).

(c) Lewis electron-dot structure for the most important resonance structures of diazomethane, CH_2N_2 (the sequence of the bonded atoms is H_2-C-N-N) showing formal charges and non-bonded electron pairs:

(d) A substance containing no ir bands between 1680 and 1780 cm^{-1} which when treated with Br_2 and NaOH followed by work-up with dilute acid gives $CHBr_3$ and 3,3-dimethylbutanoic acid.

5. (40 pts, 20 min) Free radical chlorination of 1-chloropropane with chlorine gas at 158° gives a mixture of three compounds A, B, and C, all of which have a molecular weight of 113 and the following composition: C, 31.9%; H, 5.3%; Cl, 62.8%. The nmr spectra of A, B, and C are:

A	B	C
5.3δ 1H triplet	2.8δ 2H quintet	5.0δ 1H multiplet
2.0δ 2H multiplet	1.6δ 4H triplet	3.2δ 2H doublet
1.0δ 3H triplet		1.6δ 3H doublet

Deduce the structures of A, B, and C. Clearly designate which structure is associated with each letter. If the products are formed in the ratio:
A:B:C = 22:32:46
deduce the relative reactivities of the different types of hydrogen atoms in 1-chloropropane in the reaction described.

6. (25 pts, 12 min) Suggest methods for performing the following reactions.
   (a) [benzene] ⟶ [diphenylmethane: Ph-CH$_2$-Ph]

   (b) [PhCHO] ⟶ $CH_3CHBrCHBrCHBrC_6H_5$

   (c) [toluene] ⟶ [benzoic acid with Br and NO_2 substituents: COOH, Br, NO_2]

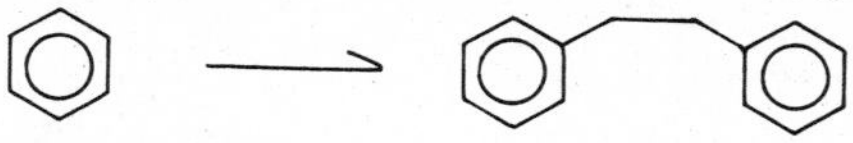

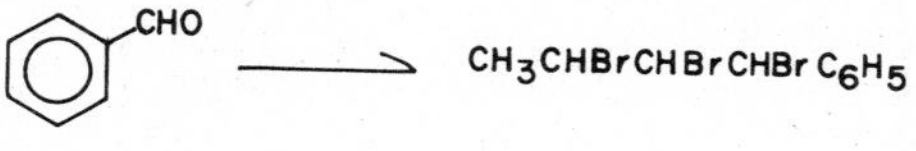

7. (25 pts, 12 min) When the R enantiomer of Z-5-bromo-3-methyl-2-hexene is treated with bromine in methanol, among the products are found various substances having the molecular formula $C_7H_{14}OBr_2$. Give projection formulas for the most likely constituents of the mixture of $C_7H_{14}OBr_2$ compounds and indicate how many fractions should be obtained on careful fractional distillation of the mixture.

8. (20 pts, 10 min) In each of the following equilibrium processes, is K

greater than one or less than one? Offer a brief explanation for each.
(a) $HCl + I^- \rightleftharpoons HI + Cl^-$
(b) $C_2H_5OH + HCO_3^- \rightleftharpoons C_2H_5O^- + CO_2 + H_2O$
(c) $C_6H_{11}OH + C_2H_5^- \rightleftharpoons C_6H_{11}O^- + C_2H_6$

(d) $\underset{HO}{\overset{H}{>}}C=CH_2 \rightleftharpoons CH_3CHO$

(e) $NH_3 + C_2H_5MgCl \rightleftharpoons C_2H_6 + Mg(NH_2)Cl$

9. (20 pts, 10 min) Explain the following:
(a) A solution of R-2-iodooctane in acetone loses all traces of optical activity a short time after some NaI is added to the solution.

(b) Only one of the following is suitable for the preparation of t-butyl methyl ether:

$$CH_3O^- + (CH_3)_3CBr \longrightarrow CH_3OC(CH_3)_3$$

$$(CH_3)_3CO^- + CH_3Br \longrightarrow CH_3OC(CH_3)_3$$

Final Examination (A)   Answer Set

1. (a) This is an example of the oxymercuration-demercuration procedure.

   [cyclopentene-cyclohexane structure] $\xrightarrow{Hg(OAc)_2, H_2O}$ [intermediate with OH and HgOAc] $\xrightarrow{NaBH_4}$ [product with OH]

   The addition occurs across the double bond with Markownikov orientation. The -OH group becomes attached to the C atom that can best accommodate a positive charge.

   (b) The first step is the protonation of the starting 2-methylpropene, giving the <u>t</u>-butyl cation, I.

   $$(CH_3)_2C=CH_2 \longrightarrow (CH_3)_3C^+$$

   The cation I is then attacked by another molecule of alkene producing cation II.

   $$(CH_3)_2C=CH_2 + (CH_3)_3\overset{+}{C} \longrightarrow (CH_3)_2\overset{+}{C}CH_2C(CH_3)_3$$

   Loss of a proton from II will give a mixture of A and B.

   A, $H_2C=C(CH_3)CH\,C(CH_3)_3$;   B, $(CH_3)_2C=CHC(CH_3)_3$

   Catalytic hydrogenation of A and B will lead to the same product C.

   C, $(CH_3)_2CHCH_2C(CH_3)_3$

   (c) $C_2H_6 + CH_3C\equiv CMgBr$

   Methylacetylene contains an acidic proton. This is abstracted by the Grignard reagent giving ethane.

   (d) $CH_3CH_2\underset{O}{\overset{\|}{C}}CH_3 + CO_2$

   (e)   I, $[(CH_3)_2CHCH(CH_3)]_3 B$ II, $(CH_3)_2CHCH(OH)CH_3$

   The hydroboration procedure results in the formation of an alcohol corresponding to what one would obtain by anti-Markownikov addition of water to the parent alkene.

   (f) $C_2H_5-CH_3$

   (g)   [m-nitroacetophenone structure with NO_2]

   (h) The action of cold aq. KMnO_4 on an alkene produces a 1,2-diol by <u>syn</u> addition. Thus a <u>meso</u> compound must be produced if the starting alkene is cis-2-butene.

   [Fischer projection:
   H_3C―|―H
   H_3C―|―H
   with OH on top and OH on bottom]

   (i) The -NO_2 group in the starting material will deactivate one of the phenyl rings. Thus, further substitution will occur in the unaffected ring.

(j) [structure: O$_2$N-C$_6$H$_4$-CH$_2$-C$_6$H$_4$-NO$_2$]

C$_6$H$_5$COCH$_3$

2. The percentage of C and H total only 73.3. Assuming the remaining 27.7% is oxygen (a usual first guess in the absence of conflicting evidence) the molecular formula of A is indicated as C$_3$H$_8$O. Since A is oxidized first to an aldehyde and then further to a carboxylic acid, it is presumably a 1° alcohol, and n-propanol is the only possibility. The oxidation sequence can thus be represented:

$$\underset{A}{C_2H_5CH_2OH} \longrightarrow C_2H_5CHO \longrightarrow C_2H_5COOH$$

One is told that A reacts with KBr in sulfuric acid to produce B. Since A is a primary alcohol this reaction is readily interpreted as S$_N$2 displacement of a protonated hydroxyl group (i.e. water) by bromide ion.

$$\underset{A}{C_2H_5CH_2OH} \xrightarrow{H^+} C_2H_5CH_2\overset{+}{O}H_2 \xrightarrow{Br^-} \underset{B}{BrCH_2C_2H_5} + H_2O$$

The remaining transformations can be interpreted as dehydrohalogenation to give C, Markownikov addition of HBr across the double bond of C, and finally hydrolysis to give 2-propanol.

$$BrCH_2C_2H_5 \longrightarrow \underset{C}{H_2C=CHCH_3} \longrightarrow \underset{D}{H_3CCH(Br)CH_3} \longrightarrow \underset{E}{CH_3CH(OH)CH_3}$$

3. Since a diketone is obtained by ozonolysis of the hydrocarbon C$_{10}$H$_{16}$ one must conclude that the hydrocarbon is cyclic. Furthermore, since the diketone is symmetrical, so must be the hydrocarbon. One is further told that the hydrocarbon contains no alkyl substituent groups. The only conceivable answer is thus:

[structure: decalin → two cyclohexanone rings]

4. (a) [Fischer projection with Br, CH$_3$, H$_3$C, Cl, H substituents; labeled R, R]

(b) [structure: C$_6$H$_5$-CH(OH)-CH$_3$]

(c) $\underset{H}{\overset{H}{>}}C=\overset{+}{N}=\overset{-}{N}: \longleftrightarrow \underset{H}{\overset{H}{>}}\overset{-}{C}-\overset{+}{N}\equiv N:$

(d) The formation of CHBr$_3$ suggests a haloform reaction; however, the original substance is obviously not a methyl ketone on account of its ir spectrum. Thus one must assume that the original substance is an alcohol that is oxidized to a methyl ketone under the conditions of the reaction. Thus one has the answer:

$$(CH_3)_3CCH_2CH(OH)CH_3$$

5. The percentage composition data lead to the formula $C_3H_6Cl_2$ for compounds A, B, and C. The nmr spectral data allow one to identify which structures are associated with the letters A, B, and C.

$$\underset{A}{CH_3CH_2CHCl_2} \qquad \underset{B}{H_2ClCH_2CH_2Cl} \qquad \underset{C}{CH_3CH(Cl)CH_2Cl}$$

It is needed to calculate the relative reactivities of the different types of hydrogen atoms in 1-chloropropane.

$$\underset{a}{CH_3}\underset{b}{CH_2}\underset{c}{CH_2Cl}$$

To do this one must recognize that the relative abundances of the isomers in the reaction product mixture are dictated by both statistical and reactivity factors. In this case the statistical factors are given simply by the numbers of hydrogens of each type. There are three type a, two type b, and two type c. Thus:

Rel. abundance of A = 2 x Reactivity of type c H atoms
Rel. abundance of B = 3 x Reactivity of type a H atoms
Rel. abundance of C = 2 x Reactivity of type b H atoms

Using the given relative abundances one obtains
22 = 2 x c reactivity
32 = 3 x a reactivity
46 = 2 x b reactivity

Thus the reactivities a:b:c of H atoms is
32/3 : 46/2 : 22/2   which is approximately   1:2:1

6. (a) There are various ways of accomplishing the synthesis. Two fairly direct ones are as follows:

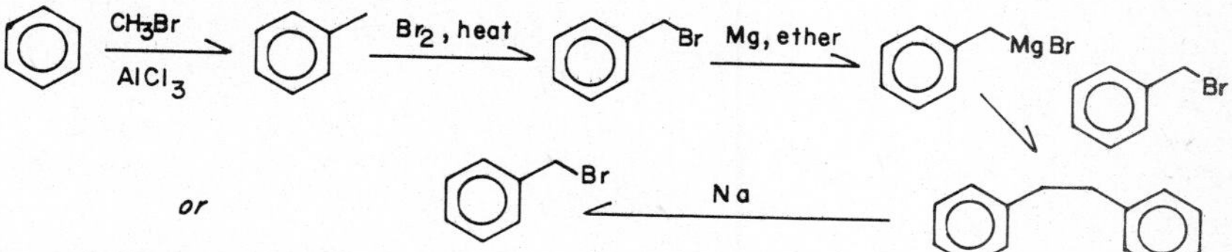

[Reaction scheme: benzene → (CH₃Br, AlCl₃) → toluene → (Br₂, heat) → benzyl bromide → (Mg, ether) → benzyl MgBr → bibenzyl, with benzyl bromide + Na also giving bibenzyl]

The Corey-House method could also be used.

$$(C_6H_5CH_2)_2CuLi + C_6H_5CH_2Br \longrightarrow C_6H_5CH_2CH_2C_6H_5$$

(b) [Reaction scheme: PhCHO → (1. C₃H₇MgI, 2. H₂O) → PhCH(OH)C₃H₇ → (acid) → PhCH=CHCH₂CH₃ → (NBS) → PhCH=CHCHBrCH₃; with Br₂/CCl₄ giving PhCHBrCHBrCHBrCH₃]

75

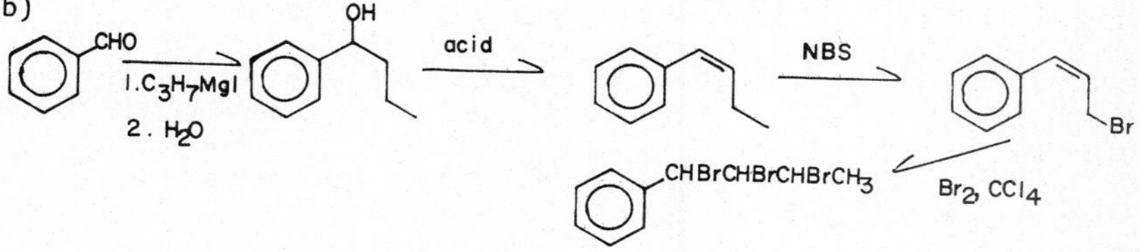

(c) [reaction scheme: benzene → Br₂, Fe → bromobenzene → hot aq. KMnO₄ → p-bromobenzoic acid → HNO₃, H₂SO₄ → 4-bromo-3-nitrobenzoic acid]

7. There are a number of stereochemical points to watch in this problem. The basic reaction is the opening of a bromonium ion by attack of methanol.

[scheme: CH₃OH + bromonium ion → −H⁺ → product with Br and OCH₃]

However, since the starting alkene is nonsymmetric, so is the intermediate bromonium ion. Accordingly, depending on which carbon atom is attacked by the methanol, a different set of products would be obtained. One must first look rather carefully, therefore, at the orientation of the reaction remembering that in such reactions it is the carbon atom that can best accommodate positive charge that is attacked. One can predict that the -OCH₃ group will attach to the third carbon atom in the molecule.

[structure: H–C(CH₃)(Br)–CH₂C(CH₃)=CHCH₃, with arrow: CH₃OH will attack here]

Other factors one must consider are the <u>anti</u> nature of the attack of methanol on the bromonium ion, and the fact that there is already one chiral carbon atom in the molecule. One must also remember to draw the bromonium ion in all possible ways compatible with its formation. Keeping all this in mind, consider the entire sequence.

[scheme showing alkene H₃C\C=C/H with G and CH₃ groups, treated with Br₂ to form two bromonium ions, attacked by CH₃OH, losing H⁺ to give products I and II]

[G = (R)-CH₃CH(Br)CH₂-]

Since the group G contains a chiral carbon atom which is specified as <u>R</u>, I and II are diastereoisomers. Two fractions were therefore obtainable on fractional distillation if attack of the methanol were exclusively at the 3° carbon atom. (How many fractions would be obtained if attack took place at both 3° and 2° carbon atoms in the bromonium ion?)

8. (a) <u>K</u> is less than one. HI is a stronger acid than HCl, so the equilibrium will favor HCl and I⁻. In general, the acids formed by the larger atoms within a particular group of the periodic table are stronger than the acids formed by the smaller atoms within the same group. This is related to the decreasing strength of the bonds to hydrogen as the size of the atom increases.

(b) Alcohols are weaker acids than carboxylic acids, thus CO_2 is not evolved when a bicarbonate is treated with an alcohol. $\underline{K}$ is less than one.
(c) Cyclohexanol is a stronger acid than ethane, so the equilibrium will favor the alkoxide ion plus ethane. $\underline{K}$ is greater than one. In general, when comparing acids formed by elements of different groups of the periodic table, the more electronegative atom (oxygen in this case) forms the stronger acid.
(d) The keto/enol tautomerism depicted strongly favors the keto form. Thus $\underline{K}$ is greater than one.
(e) Using arguments similar to those in part (c), one would predict that the reaction would proceed to the right. $\underline{K}$ is greater than one.

$$NH_3 + C_2H_5MgCl \rightleftharpoons C_2H_6 + Mg(NH_2)Cl$$
Stronger acid   Weaker acid

9. (a) When sodium iodide is added to a solution of R-2-iodooctane in acetone nucleophilic displacement by the added iodide can be expected. This frees the original iodide of the organic compound to the solution. To keep track of the two sources of iodide, label the iodide originally present in the sodium iodide with an asterisk. The nucleophilic displacement then is:

$$I^{-*} + C_6H_{13}CH(I)CH_3 \longrightarrow C_6H_{13}CH(I^*)CH_3 + I^-$$

S_N2 conditions will result in inversion of configuration:

$$H_3C-\underset{C_6H_{13}}{\overset{H}{C}}-I^{*} + I^- \longrightarrow I-\underset{C_6H_{13}}{\overset{H}{C}}-CH_3 + I^{-*}$$

Thus for every $\underline{R}$ molecule attacked by iodide, an $\underline{S}$ molecule will result. However, the displaced iodide can initiate its own nucleophilic substitution reaction resulting in inversion of configuration of another molecule. Eventually (actually in a very short period of time) a chain of inversion reactions will result in there being equal numbers of $\underline{R}$ and $\underline{S}$ molecules, i.e. a racemic mixture. Further inversions will then continue to occur, but statistics dictate that at any instant of time there will be just about equal concentrations of both enantiomers.
(b) Use of a tertiary halide will result mainly in elimination product.

$$H_3CO^- + (CH_3)_3CBr \longrightarrow (CH_3)_2C=CH_2$$

$$\cancel{\longrightarrow} H_3COC(CH_3)_3$$

Final Examination (B)   Two and One-Half Hours

1. (30 pts, 15 min) Explain the following:
   (a) When allyl bromide is refluxed with magnesium in ether appreciable amounts of a compound C_6H_{10} are produced. This compound is converted to n-hexane when treated with hydrogen in the presence of a catalyst.
   (b) When the alcohol $(CH_3)_3CCH_2OH$ is heated with acid it is converted to a mixture of two compounds, I and II, both of which have the formula C_5H_{10}. Identify the major product.
   (c) When R-3-iodo-3-methylhexane is placed in aqueous acetone solution, it is converted to racemic 3-methyl-3-hexanol.

2. (30 pts, 15 min)
   (a) Place in order of base strength (strongest first) and explain briefly.
      (i) Cl^- (ii) $^-CH_2CHO$ (iii) $(CH_3)_3CO^-$
      (iv) $CH_3CH_2O^-$ (v) CH_3CH_2OH (vi) CH_3^-
   (b) Which will undergo electrophilic sulfonation faster with concentrated sulfuric acid, bromobenzene or anisole? Explain.
   (c) Comment on the feasibility of the following syntheses:
      (i) $C_2H_5CN + I^- \longrightarrow C_2H_5I + CN^-$
      (ii)

   $$\underset{\underset{CH_3}{|}}{\overset{\overset{C_2H_5}{|}}{H_3C-C-I}} + CN^- \longrightarrow \underset{H}{\overset{H_3C}{\diagdown}}C=C\underset{CH_3}{\overset{CH_3}{\diagup}} + HCN + I^-$$

   (Dissociation constant for HCN = 4×10^{-10})

3. (20 pts, 10 min) One mole of propane was treated with two moles of chlorine under conditions that favored free radical substitution. Four dichloropropanes were isolated from the reaction mixture. The nmr spectra of these compounds are given in the following table. Deduce the structure associated with each material.

      A   δ 2.4, singlet, 6H

      B   δ 1.2, triplet, 3H;   δ 1.9, quintet, 2H
            δ 5.8, triplet, 1H

      C   δ 1.4, doublet, 3H;   δ 3.8, doublet, 2H
            δ 4.3, sextet, 1H

      D   δ 2.2, quintet, 2H;   δ 3.7, triplet, 4H

4. (20 pts, 10 min) Deduce the structure of compound A from the following information. The ir spectrum contains a strong peak at 3400 cm^{-1}. The mass spectrum contains peaks at m/e 136, 118, 107, 79, 77, 51, and 39. The nmr spectrum of A is as follows:
   δ 0.8, triplet, 3H; δ 1.6, quartet of broad peaks, 2H;
   δ 3.9, broad singlet, 1H; δ 4.3, triplet, 1H; δ 7.2, singlet, 5H

5. (15 pts, 8 min)
   (a) Given the Newman projections 1-4, pick out:
      (i) a pair of enantiomers
      (ii) a pair of diastereoisomers
      (iii) two structures that represent the same stereoisomer

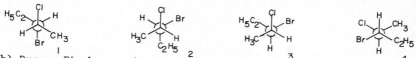

   (b) Draw a Fischer projection for the compound you selected in part (iii) of the previous problem and indicate R and S assignments.

6. (30 pts, 15 min) A reaction   A + 2B ⟶ C + D   is known to proceed via an intermediate X. The overall reaction obeys the rate law
       Rate of appearance of C = k[A][B]
   Propose a mechanism and draw a reaction coordinate diagram. Suppose that A and B start off with equal concentrations and that the initial rate is 1.6×10^{-5} M sec^{-1}. What would be the rate when 25% of A is consumed?

7. (25 pts, 12 min)
   (a) Show how 2-methyl-2-pentanol could be made using alcohols of three carbon atoms or fewer as the only organic starting materials. Any inorganic materials may be used along with the usual bench solvents.
   (b) Account for the following reaction sequence:
      (I) C_6H_{10} $\xrightarrow[H_2O]{HgSO_4, H_2SO_4}$ (II) $\xrightarrow{\text{hot aq. } KMnO_4}$ (III) (optically active)
   (c) Which of the following pathways will dominate, and why?

   [Path 1: cyclohexyl-S(+)(CH3)I⁻ + heat → cyclohexyl-S-CH3 + CH3I]
   [Path 2: → cyclohexyl-I + CH3SCH3]

8. (30 pts, 15 min) Indicate the steps by which the following synthetic conversions could be accomplished. Indicate all reagents and specify conditions. For benzene derivatives, assume that a para isomer can be separated satisfactorily from an ortho/para mixture.
   I.   cyclopentanol   ⟶   bicyclo[3.1.0] hexane
   II. tert-butanol   ⟶   2,2-dichloro-1,1-dimethylcyclopropane
   III. benzene   ⟶   p-nitrostyrene
   IV. benzene   ⟶ O_2N-C$_6H_4$-CH_2D

9. (30 pts, 15 min) Compound I, $C_5H_{10}O$, absorbs at 1715 cm^{-1}, but not above 3000 cm^{-1} in the ir. It gives a positive iodoform test and reacts with PCl_5 to give compound II, $C_5H_{10}Cl_2$. Compound II sontains in its nmr spectrum a treee proton singlet, a one proton septet, and a six proton doublet. When II is treated with alcoholic potassium hydroxide it is converted to a hydrocarbon III, C_5H_8, which absorbs near 2000 cm^{-1} in the ir.

Compound III is isomerized to IV when boiled with sodium in an inert solvent. Compound IV (but not III) reacts with methyl magnesium bromide to give compound V and methane. Acetone reacts with V to give VI (after work up with dilute aqueous acid). Compound VI takes up two molar equivalents of hydrogen under catalytic conditions to form 2,5-dimethyl-2-hexanol. Interpret the above sequence and give structures for the numbered compounds.

10. (15 pts, 8 min) Suggest explanations for the following:
    (a) Two products, both with the formula CH_3NO_2 result from the reaction of sodium nitrite solution with methyl iodide.
    (b) In dimethylformamide (1) one does not find the free rotation about the central C-N bond that one is generally accustomed to find for single bonds.

$$\begin{array}{c} H \\ \diagdown \\ O \end{array} C - \ddot{N} \begin{array}{c} CH_3 \\ \diagup \\ \diagdown \\ CH_3 \end{array}$$

(1)

Final Examination (B)   Answer Set

1. (a) When allyl bromide is refluxed in dry ether with magnesium, one can expect the corresponding allyl Grignard to form. Nucleophilic attack by this Grignard reagent on unreacted allyl bromide will give the observed product, 1,5-hexadiene

   H$_2$C=CHCH$_2$Br + H$_2$C=CHCH$_2$MgBr $\longrightarrow$ H$_2$C=CHCH$_2$CH$_2$CH=CH$_2$

   (b) (CH$_3$)$_3$CCH$_2$OH $\xrightarrow{acid}$ (CH$_3$)$_3$CCH$_2^+$ $\xrightarrow{methyl\ shift}$ (CH$_3$)$_2\overset{+}{C}$CH$_2$CH$_3$
   + H$_2$O

   $-H^+$ ↙   ↘ $-H^+$

   (CH$_3$)$_2$C=CHCH$_3$ H$_2$C=C$\begin{smallmatrix}CH_3\\C_2H_5\end{smallmatrix}$

   (c) Since R-3-iodo-3-methylhexane is a 3° halide, one can expect it to undergo reaction by the S$_N$1 route.

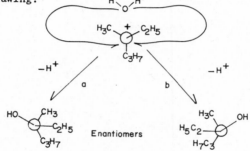

   The planar carbonium ion intermediate can then be attacked by nucleophilic reagents (water) giving a mixture of R and S products as indicated in the following drawing.

2. (a) The structures in which a negative charge is placed on carbon can be expected to be more basic than those in which negative charge is placed on the more electronegative oxygen (as long as there is not substantial resonance stabilization for the carbanionic species). Thus CH$_3^-$ and $^-$CH$_2$CHO will be the most basic. Of these two, CH$_3^-$ is the more basic because there is no resonance stabilization. Similar reasoning leads to the complete order:

   Most Basic   Least Basic

   CH$_3^-$ > $^-$CH$_2$CHO > (CH$_3$)$_3$CO$^-$ > CH$_3$CH$_2$O$^-$ > CH$_3$CH$_2$OH > Cl$^-$

   (b) Anisole will undergo electrophilic substitution reactions faster than bromobenzene because the Br atom deactivates a benzene ring toward this type of reaction. The electronegativity of bromine is such that the inductive effect of the halogen (electron withdrawing) overrides the resonance effect (electron donating). For anisole, the resonance effect predominates.

(c) (i) Since HCN is a very weak acid, its conjugate base, CN^-, is a very powerful base and nucleophile, and conversely, a very poor leaving group. Thus it would be very difficult to displace CN with I by nucleophilic displacement. HI is a very strong acid, so I^- is a weak base.

(ii) The alkyl halide is tertiary so elimination reactions are expected to be favored over substitution reactions. Since CN^- is a strong base the reaction as written seems acceptable.

3.

$H_3C\,CCl_2\,CH_3$ — 6H singlet ... A

$HCCl_2CH_2CH_3$ — 1H triplet, 3H triplet, 2H quartet of doublets ... B

$CH_3CH(Cl)CH_2Cl$ — 3H doublet, 2H doublet, 1H triplet of quartets ... C

$ClCH_2CH_2CH_2Cl$ — 2H quintet, 4H triplet ... D

4.

$C_6H_5CH(OH)CH_2CH_3$ — 5H singlet, 1H broad singlet, 1H triplet, 3H triplet, 2H quartet of doublets

5. (a) The simplest way to do this problem is to convert all the Newman projections to Fischer-type projections.

[Newman projection 1 with Cl, H_5C_2, H, Br, H, CH_3] → Fischer: H_5C_2—H (top), H_3C—H, Cl below; Br on top

[Newman projection 2 with Cl, H, Br, H_3C, C_2H_5, H] → Fischer: H_5C_2—H, H—CH_3; Br top, Cl bottom

[Newman projection 3 with Cl, H_5C_2, Br, H_3C, H, H] → Fischer: H—C_2H_5, H—CH; Br top, Cl bottom

82

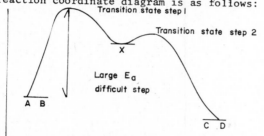

Thus 1 and 2 are diastereoisomers. So are 2 and 3. Structures 1 and 3 are enantiomers, while structures 2 and 4 are different representations of the same stereoisomer.

6. Since the rate law is Rate = $k[\underline{A}][\underline{B}]$, it follows that in the rate-determining step only one molecule of $\underline{A}$ and one of $\underline{B}$ are involved.

$$\underline{A} + \underline{B} \longrightarrow \underline{X} \quad \text{Slow step}$$

Here, $\underline{X}$ is an intermediate. Subsequently $\underline{X}$ must be converted into the products $\underline{C}$ and $\underline{D}$, and according to the stoichiometry of the overall reaction, another molecule of $\underline{B}$ is consumed. Thus one can propose the following mechanism:

Step 1 (slow) $\underline{A} + \underline{B} \longrightarrow \underline{X}$
Step 2 (fast) $\underline{X} + \underline{B} \longrightarrow \underline{C} + \underline{D}$
Overall $\underline{A} + 2\underline{B} \longrightarrow \underline{C} + \underline{D}$

A suitable reaction coordinate diagram is as follows:

Turning now to the numerical part of the problem, one is told that the initial concentrations of $\underline{A}$ and $\underline{B}$ are identical. Let this initial concentration be C_o.

$$[\underline{A}]_{initial} = [\underline{B}]_{initial} = C_o$$

Thus, 1.6×10^{-5} M sec^{-1} = $k(C_o)^2$

From which $k = \dfrac{1.6 \times 10^{-5} \text{ M sec}^{-1}}{C_o^2}$

When 25% of $\underline{A}$ is consumed its concentration will be 0.75 C_o, while that of $\underline{B}$ will be only 0.50 C_o. Thus the new set of concentration values is:
[$\underline{A}$] = 0.75 C_o [$\underline{B}$] = 0.50 C_o
Putting these values into the rate expression one obtains:

$$\text{Rate} = k(0.75\ C_o)(0.50\ C_o)$$

$$\text{Rate} = \frac{3k(C_o)^2}{8}$$

Substituting the value of k from above:

$$\text{Rate} = \frac{3 \times 1.6 \times 10^{-5}(C_o)^2}{8 \times (C_o)^2}\ \text{M sec}^{-1} = 6.0 \times 10^{-6}\ \text{M sec}^{-1}$$

7. (a)

[Structure: cyclohexane-OH] →(1. (CH₃)₂CO; 2. aq. acid)→ [structure] →(MgBr, Mg, ether)→ [structure] →(Br, PBr₃)→ [structure OH]

The acetone required for the final step in the preparation could be made from 2-propanol by dichromate oxidation.

(b) Reaction of I with $HgSO_4/H_2SO_4$ (to form II) suggests the presence of an acetylenic group. Since II reacts with bromine/sodium hydroxide solution (equivalent to hypobromite) giving a carboxylic acid one can conclude that II possesses a methylketone function. This in turn shows that I must be a terminal alkyne, R-CC-H. The R group must contain a chiral carbon atom to account for the optical activity of III. Since I contains a total of only six carbon atoms, only one structure can be drawn which contains a chiral carbon atom.

$$\underset{\underset{CH_3}{|}}{\overset{\overset{C_2H_5}{|}}{H-C-C\equiv C-H}}$$

(c) The reaction involves nucleophilic attack of I⁻ and displacement of a dialkyl sulfide. Attack can be expected to occur at the least hindered site, i.e. on a methyl group rather than on the cyclohexane ring. Pathway 1 is thus the favored route.

I⁻ + [cyclohexyl-S⁺(CH₃)₂] →(SN2 conditions)→ [cyclohexyl-SCH₃] + CH₃I

8. I.

[cyclopentanol] →(acid)→ [cyclopentene] →(CH_2I_2, Zn(Cu))→ [bicyclic structure]

II.

$(CH_3)_3COH$ →(acid)→ $(CH_3)_2C=CH_2$ →(KOH, $CHCl_3$)→ [structure with Cl Cl]

III.

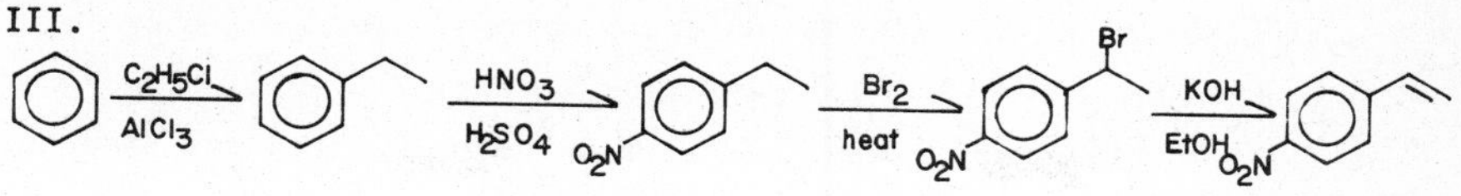

[Reaction scheme: benzene → (C₂H₅Cl/AlCl₃) → ethylbenzene → (HNO₃/H₂SO₄) → p-nitroethylbenzene → (Br₂, heat) → α-bromo-p-nitroethylbenzene → (KOH/EtOH) → p-nitrostyrene]

The double bond should be introduced after the nitration step. Why?

IV.

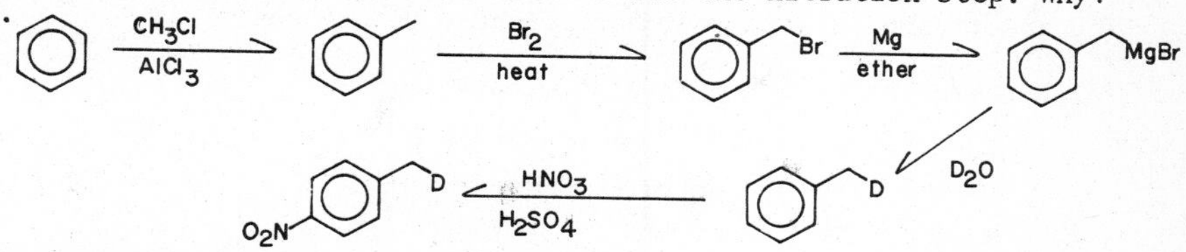

[Reaction scheme: benzene → (CH₃Cl/AlCl₃) → toluene → (Br₂, heat) → benzyl bromide → (Mg/ether) → benzylMgBr → (D₂O) → PhCH₂D; then → (HNO₃/H₂SO₄) → p-O₂N-C₆H₄-CH₂D]

The nitro group can be introduced only after all the manipulations with the Grignard reagent are complete.

9. In problems of this sort it is a good idea to summarize the information given in a flow chart.

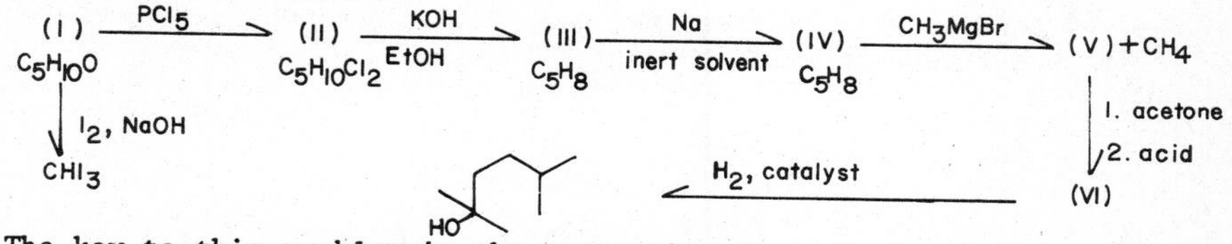

$C_5H_{10}O$ (I) $\xrightarrow{PCl_5}$ $C_5H_{10}Cl_2$ (II) $\xrightarrow{KOH, EtOH}$ C_5H_8 (III) $\xrightarrow{Na, \text{inert solvent}}$ C_5H_8 (IV) $\xrightarrow{CH_3MgBr}$ (V) + CH_4

(I) $\xrightarrow{I_2, NaOH}$ CHI_3

(V) $\xrightarrow{1.\ acetone}_{2.\ acid}$ (VI)

(VI) $\xleftarrow{H_2,\ catalyst}$ [structure shown: (CH₃)₂CHCH₂CH₂C(CH₃)₂OH]

The key to this problem is the conversion IV to V plus methane, for in order that methyl magnesium bromide be converted to methane it must react with a substance containing an acidic hydrogen atom. Since compound IV is a hydrocarbon, it must be an alkyne, since these are the only hydrocarbons previously encountered that are appreciably acidic. Thus one can formulate the conversion of IV to V as follows:

$C_3H_7CCH + CH_3MgBr \longrightarrow C_3H_7CCMgBr + CH_4$

At this point one can not be sure whether the C_3H_7- group is n-propyl or isopropyl. However, the structure of the final product shows it to be isopropyl. The conversion of V to the final product must therefore be interpreted as follows:

$(CH_3)_2CHCCMgBr \xrightarrow[2.\ acid]{1.\ acetone} (CH_3)_2CHCCC(CH_3)_2OH \xrightarrow[catalyst]{H_2} (CH_3)_2CHCH_2CH_2C(CH_3)_2OH$

Turning now to compound I, one sees that it must be a methyl ketone since it gives iodoform on treatment with NaOH/I₂. Thus its structure can be represented as: [structure: CH₃-C(=O)-CH(CH₃)₂]

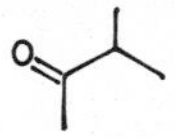

Consequently II has the structure: [structure: CCl₂(CH₃)-CH(CH₃)₂]

85

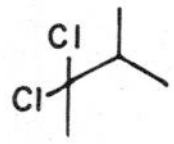

Treatment of II with KOH/EtOH could conceivably produce either:

$HCCCH(CH_3)_2$ or $H_2C=C=C(CH_3)_2$

The first of these structures has already been selected for structure IV, so the second structure must be chosen for III.

$$H_2C=C=C(CH_3)_2$$

The spectroscopic data corroborate these deductions.

10. (a) Nitrite ion is am ambident nucleophile.

$O=N-O^-$ can attack through N or O

through N gives O_2NCH_3 nitromethane
through O gives $O=NOCH_3$ methylnitrite

(b) The basic character of the nitrogen lone pair allows it to contribute to the C-N bond through the following resonance structure:

$$\underset{-O}{\overset{H}{\diagdown}}C=\overset{+}{N}(CH_3)_2$$

The C-N bond accordingly acquires some double bond character, thereby inhibiting free rotation.

Final Examination (C)   Two and One-Half Hours

1. (40 pts, 20 min) Indicate whether the following structures relate to neutral molecules or to ions (note that a species containing one positive and one negative charge is considered a neutral species), and indicate those atoms which carry charges. Also show the most important structures for 1, 3, and 4.

   [Structures 1-5 shown]

   Also, indicate whether the K_{eq} for the following reactions are greater or less than 1. Explain briefly.
   (a) $C_2H_6 + OH^- \rightleftharpoons C_2H_5^- + H_2O$
   (b) $CH_3CO_2^- + HCl \rightleftharpoons CH_3COOH + Cl^-$
   (c) $PH_3 + NH_4^+ \rightleftharpoons PH_4^+ + NH_3$
   (d) $CH_3MgBr + CH_3COOH \rightleftharpoons CH_4 + CH_3COOMgBr$

2. (30 pts, 15 min) Give structures (all possibilities) for:
   (a) A compound C_8H_{16} which, on reaction with HBr in the presence or absence of peroxides gives the same product, $C_8H_{17}Br$.
   (b) A compound $C_8H_{10}O$ which is formed by the reaction of phenyl Grignard with ethylene oxide, followed by work-up with aqueous acid.
   (c) A compound C_4H_7NO that absorbs near 2000 cm^{-1} in the ir and that has the following nmr spectrum:
       2.6δ, triplet, 2H; 3.6δ, triplet, 2H; 3.4δ, singlet, 3H
   (d) A compound C_7H_{14} that gives the same products on treatment with either hot aqueous potassium permanganate or with ozone, followed by zinc and water.

3. (25 pts, 12 min) Give the products of the following reactions:

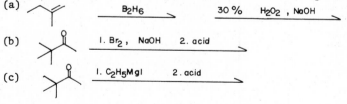

4. (20 pts, 10 min) Explain the following:
   (a) [structure: chlorocyclohexane-like bridged] is unreactive to nucleophiles by both S_N1 and S_N2 routes.

(b) When $ClCH_2CH(CH_3)CH_2CH_2NH_2$ is warmed with dilute alkali a cyclic nitrogen-containing compound is obtained and the reaction obeys first order kinetics.
(c) When ethyl iodide is treated with alcoholic KCN a mixture of two isomeric products is obtained.

5. (30 pts, 15 min) Indicate how the following synthetic conversions might be effected:
   (a) R-2-butanol ⟶ S-2-butanol
   (b) propene ⟶ 2,3-dimethyl-2-butanol
   (c)
   (d)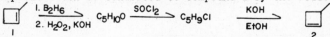

6. (25 pts, 12 min) Give structures for the following compounds:
   (a) Compound I
       Molecular weight = 109
       Analysis 33.0% C, 4.60% H, 33.0% Cl
       nmr 3H doublet(1.73δ), 1H quartet(4.47δ), 1H singlet(11.22δ)
   (b) Compounds II and III
       Compound II has the molecular formula $C_5H_{11}Br$. Reaction of II with magnesium metal in dry ether followed by treatment with methanol gives compound III. The nmr spectrum of III consists of one peak. The mass spectrum of III contains a small molecular ion at m/e 72.
   (c) Compound IV
       Formula $C_5H_{11}Br$
       nmr 6H doublet(0.80δ), 3H doublet(1.02δ), 1H multiplet(2.05δ), 1H multiplet(3.53δ).

7. (25 pts, 12 min)
   (a) Draw the most stable chair conformations of
       (i) cis-3-tert-butyl-1-methylcyclohexane
       (ii) cis-1,1,3,5-tetramethylcyclohexane
   (b) Compound 1 can be converted to compound 2 by the following sequence:

   □ —1. B_2H_6 / 2. H_2O_2, KOH→ $C_5H_{10}O$ —$SOCl_2$→ C_5H_9Cl —KOH/EtOH→ □
   1   2

   Compound 2 is the sole product. By considering the stereochemical features of the sequence of steps, deduce whether thionyl chloride ($SOCl_2$) results in replacement of -OH by -Cl with retention or inversion of configuration. Explain concisely.
   (c) Indicate which combination of organic and inorganic compounds will result in the production of the compound shown. Choose one item from each list and justify your answer in one or two sentences.

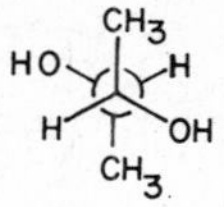

	Organics	Inorganics

```
         CH3           1-butene           (BH3)2, HO2-
    HO       H         cis-2-butene       dil. aq. KMnO4
     H      OH         trans-2-butene     Hg(OAc)2, NaBH4
         CH3           1-butyne           Hg2+, aq. acid
                       2-butyne           O3, H2O, Zn
```

8. (30 pts, 15 min)
(a) How many nmr signals does one expect for the central hydrogen atom in 3-chloropentane? Explain the origin of the peaks as clearly as possible (a clear diagram, properly annotated, will suffice).
(b) The two principal isotopes of bromine are ^{79}Br and ^{81}Br. These two isotopes are approximately equally abundant. Assuming equal abundance, estimate the relative intensities of the peaks at m/e 199, 201, and 203 in the mass spectrum of 1,3-dibromopropane. Express the intensities relative to the m/e 199 peak.

9. (25 pts, 12 min)
(a) There is a family of organic compounds known as isocyanates. They have the general formula RNCO where R is an alkyl or aryl group. These compounds may be considered derivatives of a parent compound, isocyanic acid, HNCO. Draw a Lewis structure for this compound, indicating all bonding and lone pair electrons. Further, deduce the geometry of this compound (show a clear drawing). Indicate the hybridization shcemes in operation for the nitrogen and carbon atoms.
(b) Which will be the stronger acid, cyclohexanol or phenol? Explain.
(c) Suggest a mechanism for the following reaction:

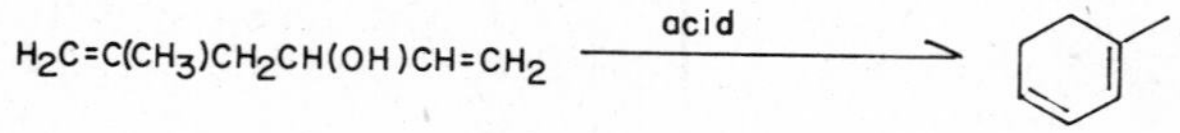

$$H_2C=C(CH_3)CH_2CH(OH)CH=CH_2 \xrightarrow{acid} \text{(toluene)}$$

Final Examination (C)   Answer Set

1. Structure 1 (a positive ion):

            Stable; every atom has an octet.

    Other structures can be drawn:

    These place positive charge on carbon, but it should be noted that the
    carbon atom bearing the positive charge has only a sextet of electrons.
    Structure 2 (a negative ion):

    :F:
    :F-B-F:
    :F:

    Structure 3 (a positive ion):

    H⟨N⟩ ⟷ H⟨N⟩+ ⟷ H⟨N⟩
    Cl H   Cl H   Cl H

    Structure 4 (a negative ion):

    H   :O: H   :O:⁻
     \\C-C⁻   ⟷   \\C=C
    H₃C   H   H₃C   H

    Structure 5 (a neutral molecule):

    Ph-N⁺=N-O:⁻ (with ring)

    Equilibria:
    (a) K less than 1; (b) K more than 1; (c) K less than 1; (d) K more than 1

2. (a) Since the alkene gives the same product with HBr whether or not per-
    oxides are present, it must be symmetrical. Thus the structure must be
    narrowed down to one of four possibilities:
        C₃H₇CH=CHC₃H₇ where C₃H₇ may be either n-propyl or isopropyl;
        cis- and trans- isomers are also possible.
    (b)

    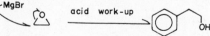

    (c)
              CH₃OCH₂CH₂CN   CN triple bond absorption near 2000 cm⁻¹ in ir
    singlet   triplet triplet

    (d) Only a tetra-alkyl substituted alkene would give the same products
    with either ozone or aqueous permanganate. Thus, only one structure can
    be proposed for the compound in question.
                H₅C₂(CH₃)C=C(CH₃)₂

3. (a) [structure: 2-methylpropan-1-ol, (CH3)2CHCH2OH... actually isobutanol-like with OH]

(b) [structure: pivalic acid, (CH3)3C-COOH]

(c) [structure: 2,3-dimethylbutan-2-ol with OH]

(d) [structure: 2-methylbutane / isopentane]

4. (a) This compound is unreactive to S_N2 displacement because approach of the nucleophile will be too sterically hindered owing to its bicyclic nature. S_N1 reaction is also impossible because a carbonium ion intermediate should be planar and this would be impossible in this case because of the bridged ring system.

(b) [structure showing Cl-CH2-CH2-CH2-CH2-NH2 → cyclic ammonium → cyclic NH + H+]

Since the amine group and a -Cl group are part of the same molecule, this is an intramolecular nucleophilic displacement reaction. Because no other molecules are involved, it must be first order.

(c) Cyanide is an ambident nucleophile, i.e. it can attack via either the nitrogen or carbon atom.

CH_3CH_2I + CN^- → $CH_3CH_2N\equiv C:^-$ an isonitrile

CH_3CH_2I + CN^- → CH_3CH_2CN a nitrile

5. (a) [Fischer projection: H3C-C(H)(OH)-C2H5] →(H3CC6H4SO2Cl)→ [H3C-C(H)(OTs)-C2H5] →(OH⁻)→ [HO-C(H)(CH3)-C2H5 inverted]

In the formation of the sulfonate ester, no bond to the chiral carbon atom is broken, so the stereochemistry is retained. Displacement of the sulfonate group with OH⁻ is an S_N2 process resulting in inversion of configuration, thereby giving the S configuration product.

(b) [synthesis scheme: 2,3-dimethyl-2-butanol ← from ketone + MeMgBr ← MgBr/Mg/ether ← Br compound ← HBr ← alkene; ketone from KMnO4 oxidation of OH (aq. acid)]

(c)

H₅C₆CH₂–CH(H)–C(C₂H₅)(H) —CH₂I₂, Zn(Cu)→ H₅C₆CH₂–C(H)=C(C₂H₅)(H) —Na, liq. NH₃→ C₆H₅CH₂CCC₂H₅

(benzene) —Br₂, heat→ (C₆H₅–Br) —(benzyl bromide with isopropyl?)→ NaCCC₂H₅ —NaNH₂→ HCCC₂H₅

(d)

HOOC(CH₂)₄COOH ←hot aq. KMnO₄— (cyclohexene) ←acid— (cyclohexanol) —LiAlH₄→ (cyclohexanone)

6. (a) The first thing to note is that the elemental analysis figures for C, H, and Cl do not add up to 100%. Under these circumstances one would normally make a first guess that the missing percentage is due to oxygen. In this case such an assumption leads to the formula $C_3H_5O_2Cl$ for I. The nmr spectrum shows a very low field (acidic) proton, suggesting a carboxylic acid, RCOOH. The only structure that is consistent with the nmr data is the following:

        H
        |
   Cl–C–CH₃
        |
      COOH

(b)

           CH₃   CH₃   CH₃
           | | |
    H₃C–C–CH₃   ←CH₃OH—   H₃C–C–CH₂MgBr   ←Mg, ether—   H₃C–C–CH₂Br
           | | |
           CH₃   CH₃   CH₃

(c)

(CH₃)₂CHCH(Br)CH₃ ← 3H doublet
  6H doublet   1H multiplet
        1H multiplet

7. (a)

(cyclohexane chair with H₃C equatorial and –C(CH₃)₃ equatorial)   (cyclohexane chair with H₃C and CH₃ both equatorial)

Both substituents equatorial   Both substituents equatorial

(b) In the last step (dehydrohalogenation by alcoholic KOH) an <u>anti</u> relationship for H and Cl is required in the compound C_5H_9Cl. Noting that only one product results from this dehydrohalogenation, one must conclude that there is only one possible structure for C_5H_9Cl

      CH₃   CH₃
       \   \
    H—⟨ ⟩—H   Not   Cl—⟨ ⟩—H
       Cl   H

The <u>cis</u> compound would have given the following mixture on reaction:

   ⟨□⟩–CH₃   and   ⟨□⟩–CH₃

Looking back to the first reaction in the sequence (hydroboration/oxidation), one must recognize that the hydroboration process results in syn addition of water across a double bond, with anti-Markownikov orientation.

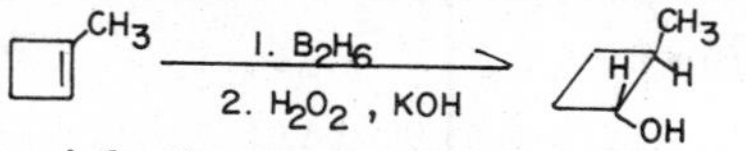

$$\text{[cyclobutane-CH}_3\text{, I]} \xrightarrow[\text{2. H}_2\text{O}_2\text{, KOH}]{\text{1. B}_2\text{H}_6} \text{[cyclobutane with CH}_3\text{, H, H, OH]}$$

Comparing this alcohol with the trans-chloro compound one must conclude that the thionyl chloride reaction occurs with retention of configuration. The entire sequence is as follows:

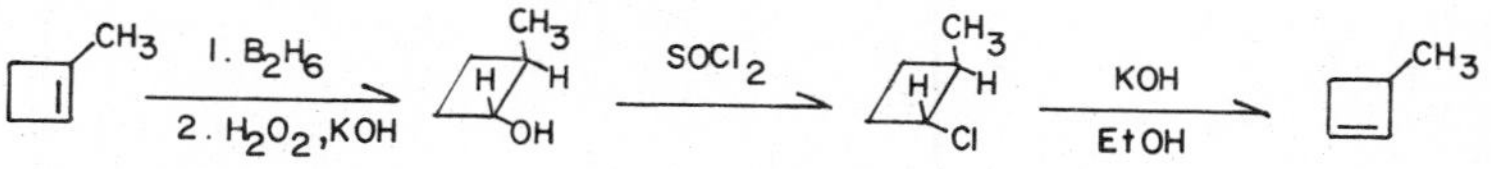

$$\text{[CH}_3\text{, I]} \xrightarrow[\text{2. H}_2\text{O}_2\text{,KOH}]{\text{1. B}_2\text{H}_6} \text{[CH}_3\text{, H, H, OH]} \xrightarrow{\text{SOCl}_2} \text{[CH}_3\text{, H, H, Cl]} \xrightarrow[\text{EtOH}]{\text{KOH}} \text{[CH}_3\text{]}$$

(c) The target compound is a meso compound.

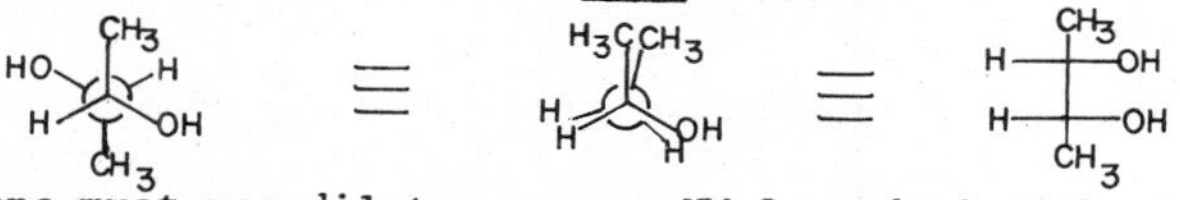

[Structures showing meso-2,3-dihydroxybutane in three equivalent representations]

Thus one must use dilute aqueous KMnO$_4$ and cis-2-butene.

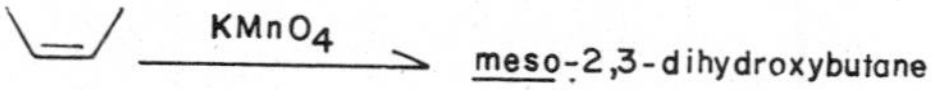

$$\text{[cis-2-butene]} \xrightarrow{\text{KMnO}_4} \text{meso-2,3-dihydroxybutane}$$

8. (a) 3-Chloropentane has the following structure:

$$\text{CH}_3\text{CH}_2\text{CH(Cl)CH}_2\text{CH}_3$$

Thus there are four equivalent hydrogen atoms on carbon atoms neighboring the central -CHCl- group. One therefore expects to see five peaks for the hydrogen atom attached to the central carbon atom, and they will arise as follows: the four adjacent proton spins will add to, or detract from, the applied magnetic field causing the central H to experience differing effective magnetic fields depending on the relative spins of these four neighboring protons. It is relatively easy to show that there are five possible combinations of these spins. Label the protons H$_A$, H$_B$, H$_C$, and H$_D$.

Combination 1. All four spins aligned to produce a net field in the same direction as the applied field. Symbolically this may be represented:

↑↑↑↑
A B C D

Combination 2. Three of the four spins are oriented together, producing a net field in the same direction as the applied field. Symbolically this may be represented:

↑↑↓↑   ↑↑↑↓   ↑↓↑↑   ↓↑↑↑
A B C D   A B C D   A B C D   A B C D

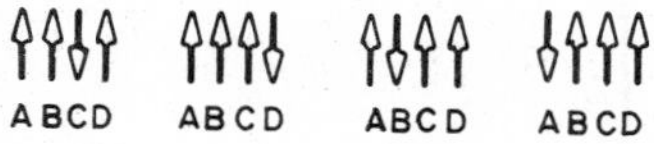

Combination 3. Two protons spin one way and two the opposite way. The fields will therefore cancel, and the central proton will experience only the applied magnetic field. Various combinations of proton spins could produce this situation.

⇡⇣⇡⇣   ⇡⇣⇣⇡   ⇡⇡⇣⇣   ⇣⇡⇡⇣   ⇣⇡⇣⇡   ⇣⇣⇡⇡
ABCD   ABCD   ABCD   ABCD   ABCD   ABCD

Combination 4. Three proton spins are aligned, but in such a way as to oppose the applied field.

⇡⇣⇣⇣   ⇣⇣⇡⇣   ⇣⇡⇣⇣   ⇣⇣⇣⇡
ABCD   ABCD   ABCD   ABCD

Combination 5. All four proton spins are aligned so as to oppose the applied field.

⇣⇣⇣⇣
ABCD

From this analysis one sees that the statistical weighting of the five combinations is 1:4:6:4:1. This will be reflected in the nmr signal for the central proton in 3-chloropentane. It will appear as a quintet whose peaks are in the intensity ratio 1:4:6:4:1.

(b) 1,3-Dibromopropane contains two bromine atoms, so one must work out the statistical chances of both these bromines being ^{79}Br, both ^{81}Br, or one of each. This is essentially the same problem as working out the relative probabilities of two tosses of a coin turning out to be two heads or two tails or one of each. There are, of course, two ways of getting "one of each" as opposed to only one way of getting two of any one specified type.

^{79}Br ^{79}Br ^{79}Br ^{81}Br ^{81}Br ^{81}Br
                        ^{81}Br ^{79}Br

Thus the m/e 199, 201, 203 peaks will have the relative intensities 1:2:1

9. (a)

H:N::C::O
  \ |/
   sp² hybridization

The sp² hybridization for the nitrogen atom necessitates trigonal geometry about that atom, while the sp hybridization for the carbon atom indicates that the N, C, and O atoms must be linear. Thus the drawing below shows the geometry of the entire molecule.

```
        120°
   H      
    \    
     N═══C═══O
    /        
         180°
```

(b) Phenol is a stronger acid than cyclohexanol because its conjugate base is resonance stabilized (the phenoxide ion). Therefore it is more readily formed.

[Resonance structures of phenoxide ion]

(c) The reaction is initiated by protonation of the hydroxyl group and loss of water, giving a carbonium ion.

[Mechanism: alcohol + H⁺ → protonated alcohol → carbocation + H₂O]

The carbonium ion is an especially stable allylic one.

[Allylic resonance structures]

One can envision the ring closure occurring via attack of the double bond π-electrons on the carbonium ion.

[Ring closure mechanism: carbocation → cyclohexenyl cation → (−H⁺) → methylcyclohexadiene]

Final Examination (D)   Two and One-Half Hours

1. (30 pts, 15 min) Give the correct answer for each part (1) - (10).
   (1) The number of hydrogen atoms in 4-methylcyclohexane.
   (2) The number of hydrogen atoms in bicyclo[3.1.0]hexane.
   (3) The molecular weight of the lowest molecular weight alkene that is chiral.
   (4) The most reactive towards Br_2/Fe: A. C_6H_5Br, B. $C_6H_5CH_2Br$, C. toluene, D. p-xylene.
   (5) The correct order of basicity: A. CH_3^-, B. $CH_3CC:^-$, C. NH_2^-.
   (6) The number of rings in phenanthrene, $C_{14}H_{10}$, if on catalytic hydrogenation with an excess of H_2 it yields a compound $C_{14}H_{24}$.
   (7) The number of equatorial methyl groups in the most stable conformation of the compound

   (8) The hybridization scheme for the central carbon atom in ketene. $H_2C=C=O$.
   (9) The number of π-electrons in the compound

   Na^+ ⬡⁻

   (10) The number of different hydrogen atoms in (a) 1,2-dichloropropane, (b) methyl t-butyl ether.

2. (25 pts, 12 min) Show the reagent or combination of reagents which will effect the following one-step conversions.

   (a) $H_3CO-\text{⬡}-SO_3H \longrightarrow H_3CO-\text{⬡}-D$

   (b) $CH_3C\equiv CC_6H_5 \longrightarrow \begin{matrix} H_3C \\ H \end{matrix} C=C \begin{matrix} H \\ C_6H_5 \end{matrix}$

   (c) $\longrightarrow H_3C \begin{matrix} Cl \\ \end{matrix} \begin{matrix} Cl \\ H \end{matrix} \begin{matrix} H \\ C_6H_5 \end{matrix}$

   (d) cyclohexene $\longrightarrow CH_3CO(CH_2)_4COOH$

   (e) benzene $\longrightarrow$ p-acetylbenzene

   (f) $C_6H_5CH_2CH=CH_2 \longrightarrow C_6H_5CH=CHCH_3$

   (g) $C_6H_5CH=CHCH_3 \longrightarrow C_6H_5CH_2CH(Br)CH_3$

   (h) $C_6H_5CH=CHCH_3 \longrightarrow C_6H_5CH=CHCH_2Br$

3. (20 pts, 10 min) Provide complete names, including (if relevant) stereo-chemical details for the following:

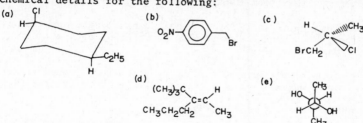

4. (16 pts, 8 min) In each part of the following set one Lewis structure is provided. In some cases one or more other significant resonance structures may be drawn. In others there may be no additional resonance structures that are reasonable. Give all significant resonance structures where appropriate showing all lone pairs and formal charges where relevant.

5. (16 pts, 8 min) Give the structures of the molecules or ions resulting from the following reactions. Give only the product resulting from the indicated processes even if these products may undergo subsequent transformation to a final product or products.

6. (16 pts, 8 min) In each of the two parts show two clear resonance structures for the most stable carbonium ion resulting from

   (a) nitration of 4-bromobiphenyl

   (b) reaction of $H_3CO-\langle\rangle-CH=CH-\langle\rangle$ with HCl

7. (64 pts, 30 min) Provide structures for the following compounds.
   (a) Compound <u>A</u>, $C_{16}H_{16}$, on ozonolysis and work up provides a single product <u>B</u>, C_8H_8O. Sodium borohydride reduction of <u>B</u> yields <u>C</u> which has the following nmr: 7.2δ, singlet, 5H; 4.2δ, singlet, 1H; 3.7δ, triplet, 2H; 2.8δ, triplet, 2H.

(b) Compound D, which contains two bromine atoms per molecule, is the single product obtained from the reaction of $C_6H_5CH=CHCH=CH_2$ with 1 mole of Br_2/CCl_4.
(c) Compounds E and F, both C_7H_{12}, and both containing 5-membered rings, are formed from the compound shown below by reaction with aq. acid.

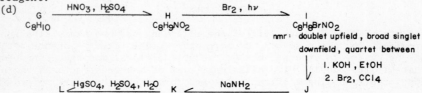

Compound E gives a single compound containing two COOH groups on treatment with hot aq. $KMnO_4$ while F gives a single diketone when treated with this reagent.

(d)
$$G \xrightarrow{HNO_3, H_2SO_4} H \xrightarrow{Br_2, h\nu} I$$
$$C_8H_{10} \qquad C_8H_9NO_2 \qquad C_8H_8BrNO_2$$
nmr: doublet upfield, broad singlet downfield, quartet between

$$\downarrow \begin{array}{l} 1.\ KOH, EtOH \\ 2.\ Br_2, CCl_4 \end{array}$$

$$L \xleftarrow{HgSO_4, H_2SO_4, H_2O} K \xleftarrow{NaNH_2} J$$

(e) Compound M, C_4H_8O, is optically active, reacts rapidly with bromine at 0°, and absorbs strongly at 3300 cm^{-1} in its ir spectrum.
(f) Compound N is the product of the following reaction:

$$\text{(structure)} \xrightarrow{H_3PO_4} N\ (C_{10}H_{18})$$

Compound N is converted to another compound, $C_{10}H_{20}$, on treatment with an excess of hydrogen under catalytic conditions. On ozonolysis N yields formaldehyde and another compound O which absorbs in the ir at 1720 cm^{-1}.
(g) Compound P, an alkene, is the major product of the following reaction:

$$C_2H_5CH(Br)CH_2\text{-cyclohexyl} \xrightarrow{KOH, EtOH} P$$

8. (20 pts, 10 min) (2R,4R)-2-Iodo-4-phenylpentane is treated with $Br_2/h\nu$ whereupon benzylic bromination occurs. A careful fractional distillation is performed following the reaction; how many fractions are obtained? Are any optically active? Name all products in each fraction complete with R,S designation. If this compound is instead treated with cyanide ion such that S_N2 conditions prevail, iodide is displaced and a cyano group is introduced. If a careful fractional distillation is performed, how many fractions will be obtained? Will any be optically active? Name the compounds in each fraction.

9. (43 pts, 20 min)
(a) If one mole each of methyl chloride and methyl bromide are stirred with one mole of KI in a suitable solvent, such as EtOH, should more CH_3Br or CH_3Cl react? Explain briefly.
(b) Predict the structure of the rearranged product (if any) in the following S_N1 reaction:

(c) Place in order of likelihood of undergoing E1 rather than E2 reaction.
   (A) $(CH_3)_2CHCH_2Br$ (B) $(CH_3)_2CHCH(Br)CH_3$ (C) $(CH_3)_2CBrCH_2CH_3$

(d) In which case would more E1 than E2 occur? Briefly explain.
   (A) $(CH_3)_3CCl$ + 0.10 $\underline{M}$ NaOH ⟶
   (B) $(CH_3)_3CCl$ + 0.050 $\underline{M}$ NH_3 ⟶

(e) Which will give the higher ratio of substitution product to elimination product? Why?
   (A) $(CH_3)_2CHCH_2Br$ + NH_3 ⟶
   (B) $(CH_3)_2CHCH_2Br$ + NH_2^- ⟶

(f) Which will be more reactive by the E2 mechanism? Explain briefly.

(A) [cyclohexane with Cl, CH$_3$, CH$_3$, and $(CH_3)_2CH$ substituents]   (B) [cyclohexane with Cl, CH$_3$, CH$_3$, and $(CH_3)_2CH$ substituents, different stereochemistry]

(g) Arrange in order of ease of dehydration.
   (A) $C_6H_5CH_2CH_2OH$ (B) $C_6H_5CH(OH)CH_3$ (C) $[C_6H_5]_2C(OH)CH_3$

(h) Arrange in order of their expected reactivity by an S_N2 reaction with bromide ion as the nucleophile.

(A) [structure with I]   (B) [structure with OH]   (C) [structure with I]

(i) Name (with $\underline{R,S}$ or $\underline{meso}$ labels) the diol resulting from $\underline{syn}$ addition of two -OH groups to $\underline{cis}$-1,2-dimethylcyclohexane.

Final Examination (D)   Answer Set

1. (1) 12; (2) 10; (3) 84; (4) D; (5) A,C,B; (6) 3; (7) 2; (8) sp; (9) 6; (10a) 4; (10b) 3.

2. (a) D_2O, D^+

   (b) Na, liq. NH_3

   (c) KO_t_-Butyl, $CHCl_3$

   (d) hot aq. $KMnO_4$

   (e) CH_3COCl, $AlCl_3$

   (f) KOH, EtOH

   (g) HBr, peroxides

   (h) N-bromosuccinimide

3. (a) _cis_-4-chloro-1-ethylcyclohexane

   (b) _p_-nitrobenzyl bromide

   (c) (_S_)-1-bromo-2-chloropropane

   (d) (_E_)-3-_t_-butyl-2-hexene

   (e) _meso_-2,3-butanediol

4. (a) [three resonance structures of benzyl cation]

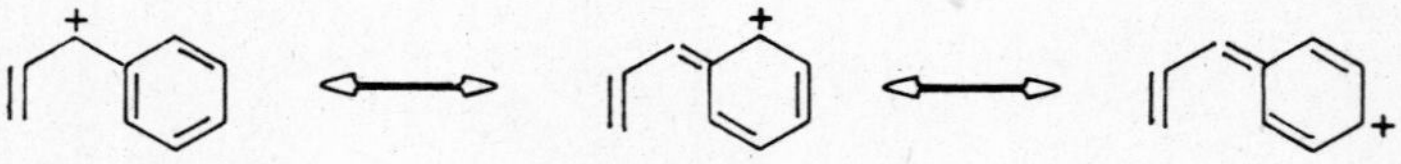

   (b) [carboxylate resonance structure]

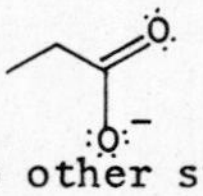

   (c) No other structures

   (d) [three resonance structures of p-methoxybenzyl chloride protonated intermediate]

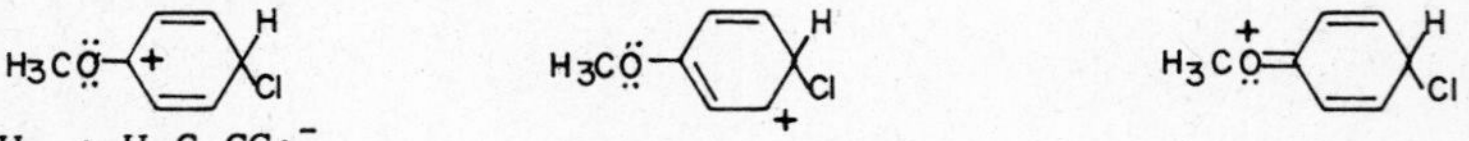

5. (a) NH_3 + $H_3C-CC:^-$

   (b) $(CH_3)_2C=\overset{+}{O}H$ + H_2O

   (c) $H_2C=CHCH_2^+$

(d) O₂N—⟨ ⟩—NH₃⁺ / Cl⁻ (with nitro on ring, ipso carbon bearing NH₃⁺ and Cl⁻)

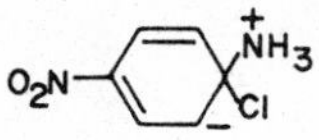

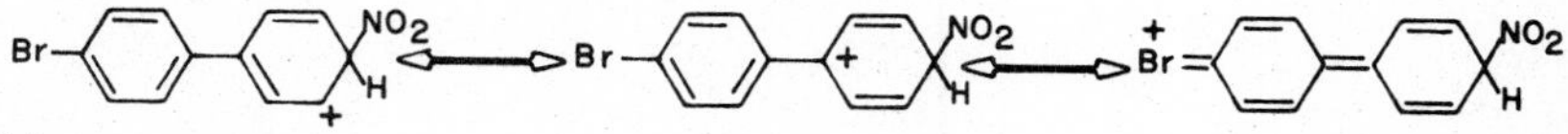

6. (a) Br—⟨ ⟩—⟨ ⟩—NO₂ resonance structures showing positive charge delocalization across biphenyl system with Br

    (b) H₃CO—⟨ ⟩—CH⁺—C₆H₅ ⟷ H₃CO⁺=⟨ ⟩=CH—C₆H₅

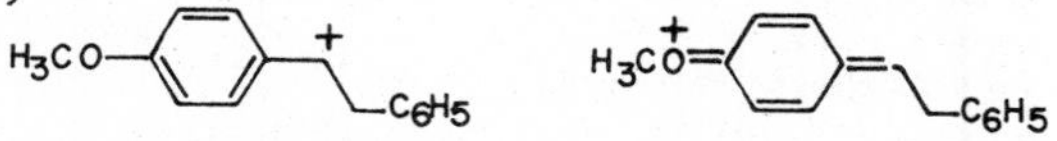

7. (a) C₆H₅CH₂CH=CHCH₂C₆H₅

    (b) C₆H₅CH₂CHO

    (c) C₆H₅CH₂OH

    (d) C₆H₅CH(Br)CH=CHCH₂Br

    (e) 1,1-dimethylcyclopent-3-ene (gem-dimethyl cyclopentene)

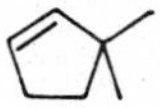

    (f) 1,2-dimethylcyclopentene

    (g) ethylbenzene

    (h) 4-ethylnitrobenzene (CH₃CH₂—C₆H₄—NO₂)

    (i) Br-CH(CH₃)—C₆H₄—NO₂

    (j) Br₂CH-CH—C₆H₄—NO₂ (1,2-dibromoethyl nitrobenzene)

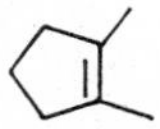

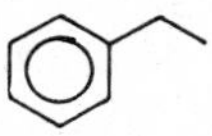

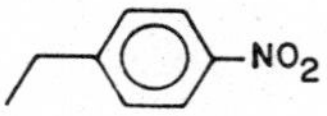

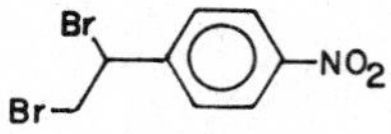

    (k) HCC—C₆H₄—NO₂

    (l) CH₃CO—C₆H₄—NO₂

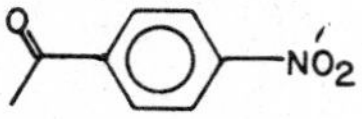

    (m) H₂C=CHCH(OH)CH₃

    (n) 1,1-dimethyl-2-isopropenylcyclopentane

101

(o) 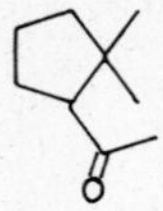 [structure: cyclopentane with gem-dimethyl and ketone substituent]

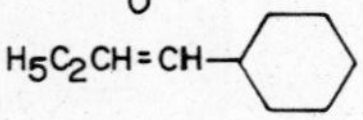

(p) H₅C₂CH=CH—⟨cyclohexane⟩   or   H₃CCH=CHCH₂—⟨cyclohexane⟩

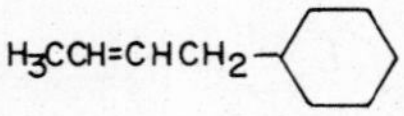

8. For the first reaction, two fractions will be obtained, both of which are optically active. The compounds present in each fraction are:
   (2R,4R)-2-iodo-4-bromo-4-phenylpentane
   and (2R,4S)-2-iodo-4-bromo-4-phenylpentane
   For the second reaction only one fraction will be obtained, it also optically active. This material is:
   (2S,4R)-2-cyano-4-phenylpentane

9. (a) CH₃Br; Br⁻ is the better leaving group.

   (b) (CH₃)₂C(Br)CH₂CH₂CH₃

   (c) C,B,A

   (d) B; use of the weaker, more dilute base makes the E2 process harder.

   (e) A; NH₂⁻ is a strong base which favors elimination.

   (f) A; H and Cl can easily assume an <u>anti</u>-relationship.

   (g) C,B,A

   (h) A,C,B

   (i) <u>meso</u>-1,2-dimethylcyclohexane-1,2-diol

Final Examination (E)   Two and One-Half Hours

1. How many tertiary hydrogen atoms are present in each molecule of the compound 5-tert-butyl-4-isopropylnonane?
   (a) 1
   (b) 2
   (c) 3
   (d) 4
   (e) more than 4

2. How many hydrogen atoms are present in each molecule of 1,3-dimethylbicyclo[1.1.0]butane?
   (a) 8
   (b) 10
   (c) 12
   (d) 14
   (e) less than 8

3. How many hydrogen atoms are present in each molecule of bicyclo[2.2.2]octa-2,5,7-triene?
   (a) 6
   (b) 8
   (c) 10
   (d) 12
   (e) less than 6

4. When bicyclo[2.2.1]hept-2-ene is treated with hot aq. KMnO$_4$ the product is
   (a) cis-1,3-cyclopentanedicarboxylic acid
   (b) 1,7-heptanedioic acid
   (c) bicyclo[2.2.1]heptane-2,3-dicarboxylic acid
   (d) cycloheptane-1,3-dicarboxylic acid
   (e) bicyclo[2.1.0]pentane-2,3-dicarboxylic acid

5. Consider the C-H bonds in (A) acetylene, (B) ethylene, and (C) ethane. In increasing length of C-H bond the order is (longest last)
   (a) A,B,C
   (b) C,B,A
   (c) A,C,B
   (d) B,C,A
   (e) C,A,B

6. Consider the compounds (A) water, (B) 1-butyne, (C) 2-butyne, and (D) cyclopropene. Which of these compounds is or are more acidic than ammonia?
   (a) A
   (b) B
   (c) B,C
   (d) A,B
   (e) B,D

7. There are two significant resonance structures for diazomethane, CH_2N_2, which is a useful methylating agent in organic chemistry. In both of these structures the C atom and the two N atoms have complete octets. The formal charge on the carbon atom in these structures is
   (a) 0,0
   (b) 0,-1
   (c) 0,1
   (d) -1,1
   (e) 1,1

8. Considering the two resonance structures (problem 7) the N-N bond in the two structures is
   (a) single, single
   (b) single, double
   (c) single triple
   (d) double, double
   (e) double, triple

9. Consider (A) F^-, (B) Cl^-, (C) OH^-, and (D) CH_3^-. The correct order of basicity (strongest base last) is
   (a) D,C,B,A
   (b) D,B,A,C
   (c) B,A,C,D
   (d) A,B,C,D
   (e) A,C,D,B

10. Consider the two C-C bonds in the allyl radical. Which of the following statements is true?
    (a) Both bonds are the same length and longer than the C-C bond in ethane.
    (b) Both bonds are the same length and longer than the C-C bond in ethene.
    (c) Both bonds are the same length and shorter than the C-C bond in acetylene.
    (d) One bond is somewhat shorter than the C-C bond in ethane and one bond is slightly longer than the C-C bond in ethene.
    (e) One bond is somewhat longer than the C-C bond in ethane and the other is slightly shorter than the C-C bond in ethene.

11. How many different monochlorinated derivatives can be made from 3,3-diethylpentane?
    (a) 1
    (b) 2
    (c) 3
    (d) 4
    (e) 5

12. If

   then

(a) A = Br, B = C₂H₅, C = F
(b) A = Br, B = F, C = C₂H₅
(c) A = F, B = Br, C = CH₃
(d) A = C₂H₅, B = Br, C = F
(e) A = F, B = C₂H₅, C = Br

13. If  then
    (a) A = Cl, B = Br, C = C₂H₅
    (b) A = Cl, B = C₂H₅, C = Br
    (c) A = C₂H₅, B = Cl, C = Br
    (d) A = Br, B = Cl, C = C₂H₅
    (e) A = C₂H₅, B = Br, C = Cl

14. How many stereoisomers can exist for 2,3,4,5-tetrachloropentane?
    (a) 2
    (b) 4
    (c) 6
    (d) 8
    (e) more than 8

15. An organic compound analyzes as follows: C, 72%; H, 12%; O, 16%. NMR shows the compound to consist of 24 H atoms per molecule. What is the molecular weight of the compound?
    (a) 50
    (b) 100
    (c) 200
    (d) 400
    (e) none of the above

16. Acetylene + 2 moles HBr yields
    (a) 1,2-dibromoethane
    (b) 1,1-dibromoethane
    (c) 1,1-dibromoethylene
    (d) 1,2-dibromoethylene
    (e) dibromoacetylene

17. Which compound is most acidic?
    (a) cyclobutane
    (b) cyclobutene
    (c) cyclopentadiene
    (d) cyclopropene
    (e) benzene

18. p-Nitroanisole + Br₂/FeBr₃ yields
    (a) 2-bromo-4-nitroanisole
    (b) 3-bromo-4-nitroanisole
    (c) no reaction

(d) 4-bromonitrobenzene
(e) bromoanisole

19. $(CH_3)_2CuLi + C_6H_5CH_2I$ yields
    (a) 4-ethyltoluene
    (b) t-butylbenzene
    (c) ethylbenzene
    (d) isopropylbenzene
    (e) styrene

20. Which will give the most exothermic reaction on hydrogenation?
    (a) 2,3-dimethyl-1-butene
    (b) 2-methyl-2-butene
    (c) trans-2-butene
    (d) cis-2-butene
    (e) 1-butene

21. Propylene + CCl_4 with peroxides yields
    (a) 1,1,1,3-tetrachlorobutane
    (b) 1,3,3,3-tetrachlorobutane
    (c) 1,1,1,2-tetrachlorobutane
    (d) 1,2,2,2-tetrachlorobutane
    (e) 1,1,4,4-tetrachlorobutane

22. 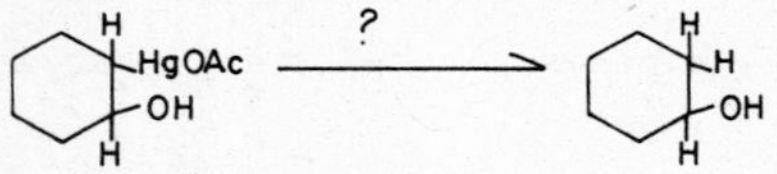 [cyclohexane with HgOAc and OH substituents → cyclohexane with H and OH substituents]

    The missing reagent is
    (a) $(BH_3)_2$
    (b) $NaBH_4$
    (c) H_2O_2
    (d) Zn, H_2O
    (e) KNH_2

23. Consider the alcohols (A) benzyl alcohol, (B) p-cyanobenzyl alcohol, and (C) p-methoxybenzyl alcohol. In order of increasing reactivity toward aq. HBr, the correct order is
    (a) A,B,C
    (b) C,B,A
    (c) C,A,B
    (d) B,A,C
    (e) A,C,B

24. Which of the following constitutes the best synthesis of 4-chlorostyrene from ethylbenzene?
    (a) (1) Cl_2, hv, (2) Cl_2, Fe, (3) KOH, EtOH
    (b) (1) Cl_2, hv, (2) KOH, EtOH, (3) Cl_2, Fe
    (c) (1) Cl_2, $FeCl_3$, (2) Cl_2, heat, (3) KOH, EtOH
    (d) (1) hot aq. $KMnO_4$, (2) CH_2N_2, (3) Cl_2, Fe
    (e) (1) Cl_2, $FeCl_3$, (2) hot aq. $KMnO_4$, (3) $NaBH_4$, aq. acid

25. Give the product of the following sequence:
    toluene + (1) HNO_3, H_2SO_4 (2) Cl_2, heat (3) C_6H_6, $AlCl_3$
    (a) p-benzylnitrobenzene
    (b) p-nitrotoluene
    (c) benzyl chloride
    (d) 4-nitrobiphenyl
    (e) p-nitrobenzyl chloride

26. Pentane can be distinguished from ethyl ether by the following test:
    (a) only ether liberates hydrogen when sodium is added
    (b) only ether dissolves in concentrated sulfuric acid
    (c) only ether gives a positive iodoform test
    (d) only ether is reduced by sodium borohydride
    (e) only ether is oxidized by aqueous dichromate

27. How many different kinds of H atoms are there present in each molecule of 2-bromobutane?
    (a) 2
    (b) 3
    (c) 4
    (d) 5
    (e) 6

28. Give the product of the following sequence:
    phenylacetylene + (1) CH_3MgBr (2) $C_6H_5CH_2Cl$ (3) Li, NH_3
    (a) trans-1,3-diphenylpropene
    (b) trans-1,4-diphenylbutene-2
    (c) 1,2-diphenylcyclopropane
    (d) p-methylstyrene
    (e) 1,2-diphenylbutane

29. Pick out the correct answers:
    (i) 1-methylcyclobutene + (1) B_2H_6 (2) HOO^- yields
        A trans-2-methylcyclobutanol
        B cis-2-methylcyclobutanol
        C 1-methylcyclobutanol
    (ii) trans-2-methyl-1-chlorocyclobutene + KOH, EtOH yields
        D 1-methylcyclobutene
        E 3-methylcyclobutene
        F a mixture of D and E
    (iii) cis-2-methyl-1-chlorocyclobutene + KOH, EtOH yields
        [Same choices as in part (ii)]
    (a) C,F,F
    (b) B,E,F
    (c) B,F,F
    (d) A,F,D
    (e) A,E,F

30. Which compound in each pair will react more readily by E2 elimination with KOH/EtOH?

(i) A 1-chlorobutane and B neopentyl chloride
(ii) C methyl propyl ether and D $CH_3CH_2CH_2OSO_2C_6H_5$
(iii) E 1-fluorobutane and F 1-iodobutane
(a) A,D,E
(b) A,D,F
(c) B,D,F
(d) B,C,E
(e) none of the above

31. Consider the structure

    In the more stable chair conformer of this compound how many methyl groups occupy equatorial positions
    (a) 0
    (b) 1
    (c) 2
    (d) 3
    (e) 4

32. 1-Bromomethylcyclobutane reacts with water and acid to yield an alkene which generates a dialdehyde on ozonolysis. The alkene is
    (a) 1-methylcyclobutene
    (b) 3-methylcyclobutene
    (c) 1,2-dimethylcyclopropene
    (d) cyclopentene
    (e) methylenecyclobutane

33. Which of the following alcohols will not react with the Lucas reagent?
    (a) cyclopentanol
    (b) 3,3-dimethyl-1-butanol
    (c) t-butanol
    (d) 2-butanol
    (e) cyclobutanol

34. Consider the reaction $CH_3Cl + CN^- \longrightarrow CH_3CN + Cl^-$: If the original concentration of CN^- is doubled, the rate of reaction, compared to the original, will be
    (a) unaffected
    (b) increased 4-fold
    (c) increased 2-fold
    (d) increased 6-fold
    (e) increased 8-fold

35. Arrange in order of rate of solvolysis, most rapid first:
    A 1-methyl-1-chlorocyclopropane
    B t-butyl chloride
    C 1-methyl-1-chlorocyclobutane
    (a) A,B,C
    (b) A,C,B

(c) B,C,A
(d) C,B,A
(e) B,A,C

36. Compound $C_xH_yO_z$ has a molecular weight of 88 and analyzes as follows: C, 68.18%; H, 13.64%. What is the value of y?
    (a) 8
    (b) 10
    (c) 11
    (d) 12
    (e) none of the above

37. Assuming that a gauche butane conformation is 0.8 kcal/mole higher than an anti conformation, what is the energy difference between the more and less stable chair conformations of 1,4-dimethylcyclohexane?
    (a) 0.8
    (b) 1.6
    (c) 2.4
    (d) 3.2
    (e) 4.0 (kcal/mole)

38. Cyclohexene + (1) CHBr₃, KO<u>t</u>-butyl, (2) Na, EtOH yields
    (a) bicyclo[4.1.0]heptane
    (b) bicyclo[3.2.0]heptane
    (c) bicyclo[3.1.1]heptane
    (d) bicyclo[2.2.1]heptane
    (e) 1,3-dibromobicyclo[1.1.0]butane

39. Propene + Cl₂/H₂O yields
    (a) 1-chloro-2-propanol
    (b) 2-chloroethanol
    (c) propyleneoxide
    (d) 2-chloro-1-propanol
    (e) 1,2-propanediol

40. <u>trans</u>-2-Methyl-1-bromocyclohexane + KOH, EtOH yields
    (a) no reaction
    (b) 1-methylcyclohexene
    (c) 3-methylcyclohexene
    (d) 4-methylcyclohexene
    (e) 1,2-dimethylcyclopentene

41. In each case pick out the more stable structure:
    (i) A benzene and B cyclooctatetraene
    (ii) C cis- and D trans-2-methyl-1-bromocyclohexane
    (iii) E   and F

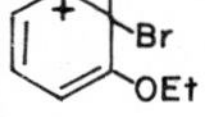

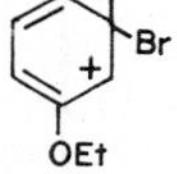

109

(a) A,D,F
(b) B,D,E
(c) A,C,F
(d) A,D,E
(e) B,C,E

42. Benzene is best converted to monodeuterobenzene by:
    (a) DCl, D_2O
    (b) Br_2, Fe; Mg, Ether; D_2O
    (c) Br_2, Fe; KOD
    (d) $CDCl_3$, $AlCl_3$; hot $KMnO_4/D_2O$
    (e) CD_3COCl, $AlCl_3$; CDI_3, KOD

43. Toluene + 2-methylpropene + HF yields
    (a) no reaction
    (b) t-butylbenzene
    (c) 1-phenyl-2-methylpropane
    (d) 4-fluoropropylbenzene
    (e) none of the above

44. Benzonitrile + $CH_3Cl/AlCl_3$ yields
    (a) no reaction
    (b) 4-methylbenzonitrile
    (c) 3-methylbenzonitrile
    (d) mixture of 2- and 4-methylbenzonitrile
    (e) none of the above

45. t-Butyl chloride + potassium t-butoxide yields
    (a) di-t-butyl ether
    (b) 2-methylpropene
    (c) no reaction
    (d) t-butyl alcohol
    (e) 2,2,3,3-tetramethylbutane

46. Which of the following constitutes the best synthesis of 2-bromo-4-nitrobenzoic acid from benzene?
    (a) (1) CH_3Cl, $AlCl_3$ (2) HNO_3, H_2SO_4 (3) $K_2Cr_2O_7$, acid (4) Br_2, Fe
    (b) (1) CH_3Cl, $AlCl_3$ (2) HNO_3, H_2SO_4 (3) Br_2, Fe (4) $K_2Cr_2O_7$, acid
    (c) (1) Br_2, Fe (2) HNO_3, H_2SO_4, (3) CH_3Cl, $AlCl_3$ (4) $K_2Cr_2O_7$, acid
    (d) (1) HNO_3, H_2SO_4 (2) Br_2, Fe (3) CH_3Cl, $AlCl_3$ (4) $K_2Cr_2O_7$, acid
    (e) (1) HNO_3, H_2SO_4 (2) H_2SO_4, fuming (3) CH_3Cl, $AlCl_3$ (4) steam (5) Br_2

47. Consider the following sequence:
    benzene + (1) C_2H_5COCl, $AlCl_3$ (2) Br_2, Fe (3) Zn(Hg), HCl (4) Br_2
    The product is:
    (a) 1-(2,3-dibromophenyl)propane
    (b) 1-(3-bromophenyl)-1-bromopropane
    (c) 1-(4-bromophenyl)-1-bromopropane
    (d) 1-(4-bromophenyl)-2-bromopropane
    (e) 1-(3-bromophenyl)-2-bromopropane

48. Which of the following reagents would most easily distinguish between benzyl alcohol and n-butylbenzene?
    (a) HNO_3, H_2SO_4
    (b) Br_2, Fe
    (c) CrO_3, H_2SO_4
    (d) Br_2, CCl_4
    (e) KOH

49. If 1-hexyne is added to a solution of ethyl magnesium bromide, a gas is evolved. What is it?
    (a) HBr
    (b) H_2
    (c) C_2H_6
    (d) Br_2
    (e) C_2H_5Br

50. If sodium is added to propanol, a gas is evolved. What is it?
    (a) C_3H_8
    (b) H_2
    (c) C_2H_4
    (d) C_3H_6
    (e) CH_4

Final Examination (E)   Answer Set

1. (c) 2. (b) 3. (b) 4. (a) 5. (a) 6. (d) 7. (b) 8. (e) 9. (c) 10. (b) 11. (b) 12. (a)
13. (b) 14. (d) 15. (c) 16. (b) 17. (c) 18. (a) 19. (c) 20. (e) 21. (a) 22. (b) 23. (c)
24. (c) 25. (a) 26. (b) 27. (d) 28. (a) 29. (e) 30. (b) 31. (d) 32. (d) 33. (b) 34. (c)
35. (c) 36. (d) 37. (d) 38. (a) 39. (d) 40. (c) 41. (d) 42. (b) 43. (b) 44. (a) 45. (b)
46. (b) 47. (b) 48. (c) 49. (c) 50. (b)

SECOND SEMESTER

First Examination (A)   One Hour

1. (25 pts, 16 min) Give the best set of reagents for each of the following conversions I-V. Please note that more than one step may be required for each conversion.

2. (25 pts, 12 min) Give the detailed mechanism for the overall conversion indicated below. The reagents shown are the only ones involved.

    $H_3C\underset{O}{\overset{\parallel}{C}}O$-◯ $\xrightarrow{H_2SO_4, H_2O}$ CH_3COOH + ◯-OH

3. (25 pts, 17 min) Give the correct structure for the compounds I-IV. Indicate proper stereochemistry where appropriate.

    (1) $C_7H_{14}O_3$ $\xrightarrow[2.\ CH_3I,\ heat]{1.\ KOH,\ CS_2}$ (II) $C_7H_{12}O_2$ $\xrightarrow[2.\ H_2O_2,\ H_2O]{1.\ O_3}$ (III) + HCOOH $C_6H_{10}O_4$

    ↓ CrO_3, pyridine   ↓ H_3O^+

    (V)   (IV)

    ↓ $LiAlH_4$, Et_2O ↓ heat
    (work-up with dil. HCl)

    (VI) + CH_3CH_2OH $CH_3CH_2COOH + CO_2$
    $C_5H_{12}O_2$

4. (25 pts, 16 min) Beginning with cyclohexanone and using any other reagents and solvents required give reasonable syntheses for each of the following compounds. Show reagents for each step of the synthetic route.

    (a) ◯ (cyclohexene)   (b) cyclohexyl-C(OH)-cyclohexyl

    (c) ◯-CH_3 (1-methylcyclohexene)   (d) $HOOC(CH_2)_4COOH$

First Examination (A)   Answer Set

1. I. [cyclohexane with CH₃ and OCH₃] ⟵ 1. CH₃OH, Hg(O₂CCF₃)₂ ; 2. NaBH₄   [cyclohexene with CH₃]

If one were to use a Williamson synthesis here, one would first need to prepare the tertiary alcohol and use it with methyl iodide. The reverse would result in an elimination reaction. Since the ether may be formed as readily as the alcohol, the alkoxymercuration-demercuration procedure shown is the preferred route.

II. [cyclohexane with CH₃ and two D] ⟵ 1. H₂NNH₂ ; 2. KOD, D₂O [cyclohexanone with CH₃] ⟵ KMnO₄ [cyclohexane with CH₃ and OH] ⟵ 1. B₂H₆ ; 2. H₂O₂ [cyclohexene with CH₃]

Againg, working backwards from the target molecule is the most feasible approach to a solution. Two hydrogen (deuterium) atoms must be added specifically to a particular carbon. Available structures do not permit the use of exchange reactions. This leaves a specific reduction of a carbonyl. Presumably a metal-acid method (Clemmensen reduction with Zn/DCl) could be used, but the Wolff-Kishner reaction shown is the preferred approach. A carbonyl group in this type of system is necessarily generated by oxidation of a hydroxyl. Permanganate is shown as the oxidizing agent, although others might be used. There remains the introduction of the hydroxyl. Hydroboration-oxidation is the method of choice as the other approaches would tend toward the tertiary alcohol. The stereochemistry of the borane oxidation is not critical here; the oxygenated carbon is made planar in the next step.

III. [cyclohexane with CH₃ and OH, stereochem] ⟵ 1. B₂H₆ ; 2. H₂O₂ [cyclohexene with CH₃]

For this conversion the reaction (hydroboration-oxidation) discussed above gives proper orientation and stereochemistry. Here the stereochemistry is critical; the alcohol has the same configuration as the borane intermediate formed by syn-addition to the alkene.

IV. [cyclohexane with CH₃ and Cl] ⟵ SOCl₂, heat [cyclohexane with CH₃ and OH]

Chlorine substitution for hydroxyl with retention of configuration is required. Knowledge of the reaction mechanism is useful for devising a synthesis. The use of thionyl chloride (or phosgene) in the absence of halide ion (as would be generated if an amine were used in the reaction medium) yields an intermediate ester-acid chloride that decomposes by an intimate ion pair mechanism upon thermolysis. Thus the desired stereochemistry is achieved. We have met this process before.

V.

[Structure: trans-1,2-dihydroxy-3-methylcyclohexane with OH groups shown with wedge/dash bonds] ⇐ HCOOH, H$_2$O$_2$ ⇐ [1-methylcyclohexene]

Hydroxylation has been effected here in a <u>trans</u>-manner. A permanganate or osmium tetroxide reaction would generate <u>cis</u> product. Thus the route should proceed through an epoxide with hydrolysis. This is accomplished most simply by using formic acid and hydrogen peroxide.

2.

$H_3C-\overset{O}{\underset{\|}{C}}-O-$[cyclohexyl] $\underset{}{\overset{[H^+]}{\rightleftharpoons}}$ [cyclohexyl]$-O-\overset{+OH}{\underset{\|}{C}}-CH_3$ $\overset{H_2O}{\rightleftharpoons}$ [cyclohexyl]$-O-\overset{OH}{\underset{+OH_2}{C}}-CH_3$

$\Updownarrow H_2O$

[cyclohexyl]$-OH$ + $H_3C-\overset{OH}{\underset{OH}{C+}}$ $\overset{H_2O}{\rightleftharpoons}$ CH$_3$COOH   ⇌   [cyclohexyl]$-O-\overset{+OH}{\underset{H}{C}}-\overset{CCH_3}{\underset{OH}{}}$ $\overset{[H^+]}{\rightleftharpoons}$ [cyclohexyl]$-O-\overset{OH}{\underset{OH}{C}}-CH_3$

This mechanism is one that most students might commit to memory. That is really unnecessary, however, since a rational consideration of acid-base reactions will ultimately lead to the answer.

First, one must consider the acid to be in reaction with some electron-rich site, two of which are in the ester (the carbonyl oxygen and the esteric oxygen). The carbonyl oxygen constitutes the better receptor of a proton in an equilibrium process, as one can see by comparing the two possible protonated species. The structure for protonation of the carbonyl oxygen is stabilized by charge delocalization (resonance}.
The protonated ester so-formed constitutes an electron deficient species and is susceptible to attack by a nucleophile. The one present in great amount is water, which adds at the carboxylate carbon. At this point, no further bond formation can occur at carbon, and only one bond, that from the proton to the trivalent oxygen atom, may be broken without involving charge separation. Reprotonation could occur at either of the two hydroxyls now present or at the alkoxide oxygen. The first possibility would be a simple reversal of the reactions above. The latter would lead to product through a neutral species (cyclohexanol) splitting off and leaving a resonance-stabilized cation, the protonated carboxylic acid.

3.

CH$_3$CH(CO$_2$C$_2$H$_5$)(CH$_2$CH$_2$OH)   CH$_3$CH(CO$_2$C$_2$H$_5$)(CH=CH$_2$)   CH$_3$CH(CO$_2$C$_2$H$_5$)(CO$_2$H)

   I   II   III

$$\text{CH}_3\text{CH}\begin{smallmatrix}\diagup\text{CO}_2\text{H}\\ \diagdown\text{CO}_2\text{H}\end{smallmatrix} \qquad \text{CH}_3\text{CH}\begin{smallmatrix}\diagup\text{CO}_2\text{C}_2\text{H}_5\\ \diagdown\text{CH}_2\text{CHO}\end{smallmatrix} \qquad \text{CH}_3\text{CH}\begin{smallmatrix}\diagup\text{CH}_2\text{OH}\\ \diagdown\text{CH}_2\text{CH}_2\text{OH}\end{smallmatrix}$$

$$\text{IV} \qquad\qquad\qquad \text{V} \qquad\qquad\qquad \text{VI}$$

One should approach this type of problem from the point at which the greatest amount of structural information is given. NOTE: one should not start by assuming a structure for the first compound; this usually leads to a completely wrong answer. In the problem at hand the most structural information is to be found in the reaction of IV, which yields propionic acid and carbon dioxide. This is a simple decarboxylation reaction that occurs readily only with geminal dicarboxylic acids. Thus IV must be α-methylmalonic acid as shown. Now one has further information with which to work toward the other structures. Compound IV is generated along with ethanol by the hydrolysis of III. Thus III must be a mono-ethyl ester of IV. A di-ester would not fit the formula given. Compound III is generated along with formic acid by an ozonolysis reaction. From the products and the formula of II one concludes that II is the terminal alkene as indicated. Compound II results from the overall dehydration (by the Chugaev method) of an alcohol; that this alcohol I is primary rather than secondary is indicated by the formation of a single alkene. The problem may now be considered from a more conventional (for the student) point of view, predicting products, given the reactants. The reaction conditions for the generation of V will allow primary alcohol oxidation to stop at the aldehyde stage. The use of lithium aluminum hydride with V will cleave the ester linkage as well as perform the reductions, yielding the diol VI.

4. As in similar preceding problems, the approach here is to consider the target molecule and the simple methods available for the generation of the functional groups present. Limitations are placed on reasonable methods by the structure of the available starting material - in this situation a carbonyl compound.

(a) cyclohexene ⇌ (H₃O⁺) cyclohexanol (OH) ⇌ (LiAlH₄, Et₂O or NaBH₄, EtOH or H₂, Pd/C) cyclohexanone

The most common routes for alkene formation are by elimination reactions. The applicable one here is dehydration as an oxygen is present in the starting material. Simple acid-catalyzed dehydration is feasible as no significant stereochemical problems are involved. More sophisticated methods such as acetate or xanthate pyrolysis, while acceptable, are not necessary. The problem then reduces to one of generating the hydroxyl group from the carbonyl, for which a variety of reagents are available.

(b) dicyclohexyl-OH ⇐ cyclohexanone + cyclohexyl-MgBr (1. PBr₃, 2. Mg, Et₂O) cyclohexanol ⇐ (a) above cyclohexanone

The fundamental problem to be faced here is one of carbon-carbon bond
formation. A tertiary hydroxyl remains at the site where this bond
formation has occurred. The obvious route for this is a Grignard reagent
in reaction with the available carbonyl compound. The Grignard reagent,
as it has the carbon skeleton of the available carbonyl compound, may
be prepared by simple reduction, halide formation, and reaction with
magnesium. Several methods for the generation of the halide from the
alcohol are available.

(c) [cyclohexene with CH₃] →(H₃O⁺)→ [cyclohexane with OH, CH₃] →(CH₃MgI)→ [cyclohexanone]

In the overall generation of this target molecule the involvement of two
fundamental reaction types is necessary: the formation of a new carbon-
carbon bond and alkene generation. Momentary consideration of ending the
route with alkylation of a vinylic position leads one to the conclusion
that the alternative sequence is preferred, thus the route shown. Alkene
generation by an elimination (dehydration) reaction is the obvious choice
here with the oxygen present in the available reagent. The hydroxyl group
is generated in alkylation with a Grignard (or alkyl lithium) reagent.

(d) HO₂C(CH₂)₄CO₂H ←(KMnO₄)— [cyclohexene] ←(a) above— [cyclohexanone]

Working back from the target molecule is extremely helpful. The ultimate
product is symmetrical, functionalized similarly at either end, and of
the same carbon number as the available reagent. One should also note
that cleavage of a carbon-carbon linkage is required. For a symmetrical
cleavage of this type an unsaturated (alkene) linkage is required,
accomplished by permanganate oxidation through the diol. The required
alkene has been generated in part (a).

First Examination (B)   One Hour

1. (25 pts, 13 min) Consider the acid catalyzed formation of propyl benzoate from benzoic acid and 1-propanol. Show all steps in the mechanism of this reaction and in a brief statement (two sentences or less) tell how this reaction may be driven to completion.

2. (25 pts, 15 min) Supply reagents and conditions for the series of conversions indicated A-E below. More than one step may be required for each conversion.

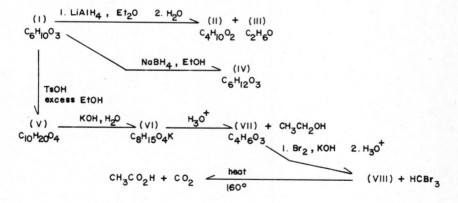

3. (25 pts, 15 min) Starting with 1-hexene, give details of the reactions (reagents and conditions) needed to convert it into each of the following compounds. Any inorganic reagents and organics of three carbons or less may be used as reactants.

   (a) heptanoic acid   (b) 3-methyl-3-heptanol
   (c) ethyl-2-hexyl ether   (d) hexanal

4. (25 pts, 17 min) Assign structures to the compounds I-VIII. Show all stereochemical details of the structures where appropriate.

$$(I)\ C_6H_{10}O_3 \xrightarrow{\text{1. LiAlH}_4,\ Et_2O\quad 2.\ H_2O} (II) + (III)\ \ C_4H_{10}O_2\ \ C_2H_6O$$

$$(I) \xrightarrow{\text{NaBH}_4,\ EtOH} (IV)\ C_6H_{12}O_3$$

$$(I) \xrightarrow[\text{excess EtOH}]{\text{TsOH}} (V)\ C_{10}H_{20}O_4 \xrightarrow{\text{KOH, H}_2O} (VI)\ C_8H_{15}O_4K \xrightarrow{H_3O^+} (VII) + CH_3CH_2OH\ \ C_4H_6O_3$$

$$(VII) \xrightarrow{\text{heat}\ 160°} CH_3CO_2H + CO_2$$

$$(VII) \xrightarrow{\text{1. Br}_2,\ KOH\quad 2.\ H_3O^+} (VIII) + HCBr_3$$

117

First Examination (B) Answer Set

1. The mechanism is simply the reverse of the one considered previously in the immediately prior set of problems. In detail this is:

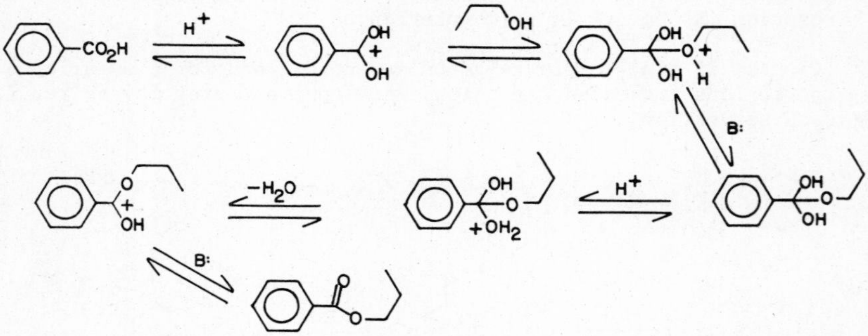

[Mechanism scheme showing conversion of PhCO₂H through protonated intermediates to the ester, with steps involving H⁺, propanol OH, loss of H₂O, and deprotonation by B:]

The species B: acting as a proton acceptor may be the anion from the acid catalyst, benzoic acid, propanol, or even water. Driving the reaction to completion may be accomplished by the removal of water from the reaction vessel as it is formed, thus displacing the equilibrium toward the ester product.

2. (A)

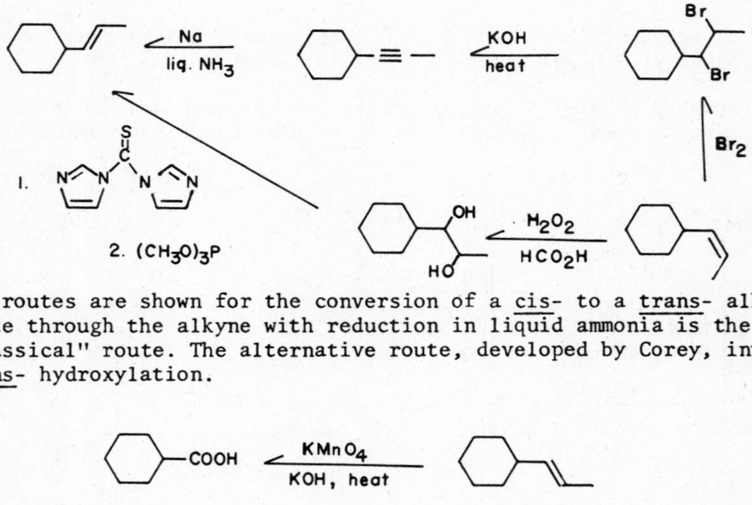

[Scheme: cyclohexyl-CH=CH₂ (cis) → via Na/liq. NH₃ ← cyclohexyl-C≡C- ← KOH/heat ← cyclohexyl-CHBr-CHBr- ← Br₂ ← trans-alkene product. Alternative Corey route: 1. thiocarbonyldiimidazole, 2. (CH₃O)₃P applied to the diol (cyclohexyl-CH(OH)-CH(OH)-) formed via H₂O₂/HCO₂H.]

Two routes are shown for the conversion of a cis- to a trans- alkene. The route through the alkyne with reduction in liquid ammonia is the "classical" route. The alternative route, developed by Corey, involves trans- hydroxylation.

(B)

[Scheme: cyclohexyl-COOH →(KMnO₄, KOH, heat)→ cyclohexyl-CH=CH-]

Other oxidizing agents might be considered here, such as ozone.

118

(C) [cyclohexyl-COCl] ←—PCl₅—— [cyclohexyl-COOH]

(D) [cyclohexyl-C(O)CH₃] ←—(CH₃)₂Cd—— [cyclohexyl-COCl]

Use of the cadmium reagent allows the reaction to be stopped at the ketone product.

(E) [cyclohexyl-CH₂CH₃] ←—Zn(Hg), HCl—— [cyclohexyl-C(O)CH₃]

A reduction technique must be used which removes the oxygen and does not simply convert the carbonyl to an alcohol. Aside from the Wolff-Kishner, other routes are unnecessarily long.

3. (a) [hexyl-COOH] ←(1. CO₂ / 2. H₂O)— [hexyl-MgBr] ←—Mg, Et₂O—— [hexyl-Br]
   ↓ H₃O⁺
   [hexyl-CN] ←—NaCN, DMSO——
                                    ←—PBr₃—— [hexyl-OH] ←(1. B₂H₆ / 2. H₂O₂)— [hexene]
                                    ←—HBr (free radical)——

Two routes are shown proceeding from the 1-bromohexane. Methods are also shown for its generation. The hydroboration-oxidation route may be preferable experimentally. Longer routes should be avoided. One should recognize at the beginning the need for the formation of a carbon-carbon bond and the addition of a single atom of carbon.

(b) [tertiary alcohol structure] ←(1. EtMgBr, Et₂O / 2. H₂O)— [ketone]
                                                                ↑ KMnO₄, KOH
    [alkene] —(1. Hg(OAc)₂, H₂O / 2. NaBH₄)→ [secondary alcohol]

This is simply the generation of a tertiary alcohol by Grignard reaction with a ketone. Hydration of the alkene by the method shown is preferred because the possibility of rearrangement is minimized. One should recognize that the target molecule requires the formation of a new carbon-carbon bond at the internal position of the starting alkene.

(c)

[Scheme showing synthesis:
- Starting material (2-hexyl ethyl ether) ← EtBr + sodium alkoxide (ONa) ← NaH from alcohol
- Alkene → 1. Hg(O₂CCF₃)₂, EtOH; 2. NaBH₄ → ether product
- Alkene → 1. Hg(OAc)₂, H₂O; 2. NaBH₄ → 2-hexanol]

Overall, two carbon-oxygen bonds must be formed in this synthesis, one at the internal position of the olefinic linkage of the starting alkene. The alternative routes shown are direct solvomercuration-reduction and the classical Williamson route. For the latter, ethoxide reaction with 2-bromohexane might be expected to yield significant elimination product.

(d) R–CHO $\xrightarrow{\text{1. CrO}_3,\ \text{Ac}_2\text{O};\ \text{2. H}_3\text{O}^+}$ R–OH

1-pentene → 1. B₂H₆; 2. H₂O₂ → pentanol

For this conversion one should recognize that only a new carbon-oxygen bond need be formed. In the first step the <u>primary</u> alcohol must be generated specifically. Oxidation of this primary alcohol must then be stopped at the aldehyde stage. We have shown one route which allows this.

4.

I — methyl 3-oxobutanoate (methyl acetoacetate)

II — 1,3-diol

III — ketal-protected alcohol

IV — methyl 3-hydroxybutanoate

V — ketal-protected ester

VI — ketal-protected potassium carboxylate (COOK)

VII — acetoacetic acid (CH₃COCH₂COOH)

VIII — HOOC–CH₂–COOH (malonic acid)

The critical aspect of solving this type of problem is knowing where to start. The point of greatest information is again at the end with all the products (acetic acid and carbon dioxide) being given. Important chemical points here are selectivity of hydride reducing agents, ketal formation for carbonyl protection, oxidation, and decarboxylation.

First Examination (C)   One Hour

1. (25 pts, 13 min) The acid catalyzed hydrolysis of benzonitrile yields benzoic acid and ammonium ion. Show all steps in the mechanism of this reaction and explain briefly (two sentences or less) what the driving force is that completes the reaction under these conditions.

2. (25 pts, 16 min) Supply reagents and conditions for the conversions I-V. More than one step may be required for each conversion.

3. (25 pts, 13 min) Starting with cyclohexanol, give details of reactions needed to convert it into each of the following compounds. One may use as reactants any inorganic compounds and organics of three carbons or less.

   (a) cyclohexane-COOH
   (b) cyclohexene-COOH
   (c) HOOC-CH₂CH₂CH₂-COOH
   (d) cyclohexyl isopropyl ether

4. (25 pts, 18 min) Assign structures to compounds A-H. Show all stereochemical details of the structures.

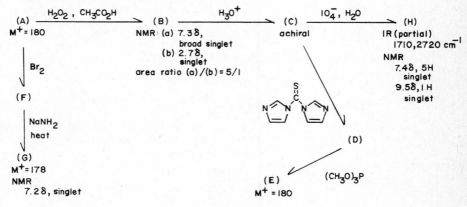

First Examination (C) Answer Set

1.

[Mechanism scheme: PhCN + H₃O⁺ ⇌ PhC≡N⁺H ↔ PhC⁺=NH ⇌ (H₂O) PhC(=NH)OH₂⁺ ⇌ (H₂O) PhC(OH)=NH₂⁺ ↔ PhC(OH)−NH₂ (+OH₂) ⇌ PhC(OH)(NH₂)⁺... → PhC(OH)−N⁺H₃ −NH₃ → PhC⁺(OH) ⇌ (H₂O) PhCOOH]

In a rationalization of product formation with acid catalysis, one looks for an electron-rich site for protonation, in this case the nitrogen. Protonation then generates an electron-deficient site which is capable of reacting with a base, in this case water. This repetition of base and acid functioning of the organic species continues until products are obtained. The fact that ammonia is protonated by the strong acid present in excess, and is virtually removed from the system by its conversion to ammonium ion, explains the driving force for the reaction.

2. (I) $KMnO_4$

   (II) CH_3OH, H^+ or CH_2N_2 or $(CH_3O)_2SO_4$, $NaOH$

   (III)   1. $EtOH$, H^+ 2. $LiAlH_4$, Et_2O 3. H_3O^+

   (IV)   1. H_2NNH_2,   2. ⁻OH, H_2O or $Sn(Hg)$, HCl

   (V)   1. $KMnO_4$,   2. KOH, H_2O, Br_2

The most complex sequence is III. Protection of the carbonyl group is necessary first, and it is accomplished by ketal formation. Reduction may then be performed followed by ketal hydrolysis.

3. (a) [cyclohexane-COOH] ⇌ (1. CO_2 / 2. H_3O^+) [cyclohexane-MgBr] ⇌ (Mg, Et_2O) [cyclohexane-Br] ⇌ (PBr_3) [cyclohexane-OH]

The main points to be noted on first observation are that an oxygen-carbon bond is broken and a new carbon-carbon bond is generated.

(b) [cyclohexene-COOH] →[KOC(CH₃)₃] [cyclohexane with Br and COOH] →[PBr₃, Br₂] [cyclohexane-COOH] →[see (a) above] [cyclohexanol]

For this relatively simple conversion students invariably invent quite long and devious routes. Having done part (a) one should already be aware of a method of generating the carboxyl group. There remains the introduction of the olefinic linkage. This is commonly accomplished by an elimination reaction (here, dehydrohalogenation), and one must recognize the route that places a halogen at the α-position of the carboxylic acid, the Hell-Volhard-Zelinsky reaction.

(c) HOOC~~~COOH →[KMnO₄ / KOH, heat] [cyclohexene] ←[H₃O⁺] [cyclohexanol]

The student should recognize two points immediately upon looking at the problem. It is necessary to cleave some carbon-carbon linkage of the starting material, and the product bears the same number of carbons as the prescribed starting material. Thus it is necessary simply to cleave the ring oxidatively after preparing the cycloalkene by dehydration.

(d) [cyclohexyl isopropyl ether] →[1.(CH₃)₂CHOH / Hg(O₂CCF₃)₂ / 2. NaBH₄] [cyclohexene] →[1.CH₃MgI / 2. H₃O⁺] [cyclohexanone] →[KMnO₄] [cyclohexanol]

It must be realized that in the overall process not only is an ether linkage to be formed, but also a new carbon-carbon bond at the site of the ether linkage. Thus two fundamental processes must be incorporated into the synthesis. For a tertiary-secondary ether the solvomercuration-reduction method is definitely superior to other methods since these others would lead to extensive elimination.

4.

A: trans-stilbene

B: trans-stilbene oxide (H₅C₆ and H on epoxide)

C: (HO)(H₅C₆)C-C(C₆H₅)(OH) stereochemistry shown

D: cyclic thiocarbonate with H₅C₆ groups

E: cis-stilbene

F: H₅C₆-CHBr-CHBr-C₆H₅ stereochemistry shown

123

G   H

(Structures: G = diphenylacetylene (Ph-C≡C-Ph); H = benzaldehyde (Ph-CHO))

The fundamental reactions and overall course of the problem are probably familiar. Conversion of A to E is an isomerization, specifically an isomerization about an olefinic linkage. The real problem is determining the specific geometries of A and E. Here the structural clue is found in the center of the scheme, at C. This compound is incapable of resolution into optically active materials, so it is a *meso* compound, having an internal plane of symmetry through the carbon-carbon bond of the vicinal diol linkage (as is readily deduced from the reactions involved). This allows the stereochemistry of A and E to be stated; *trans*- hydroxylation of a symmetrical *trans*-alkene yields a *meso* product.

First Examination (D)   One Hour

1. (25 pts, 11 min) Thionyl chloride may be used to generate alkyl chlorides from alcohols with either retention or inversion of configuration about the carbon originally bearing the hydroxyl. Give the reaction conditions necessary for each type of conversion and the mechanism for each type of reaction.

2. (25 pts, 13 min) Predict the correct organic product(s) in each of the following reactions.

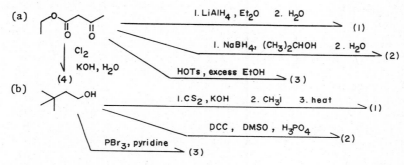

3. (25 pts, 18 min) Give the best possible synthetic route for each of the following compounds using the indicated starting material as one of the reagents and any other reagents deemed necessary. All reagents should be indicated clearly and proper stereochemistry shown.

(a)

(b) meso-2,3-dihydroxybutane from 2-butyne
(c) 3-methylbutyric acid from 2-methyl-1-propanol
(d) butyl phenyl ketone from benzyl alcohol

4. (25 pts, 18 min) Give the correct structure for each of the compounds indicated A-F.

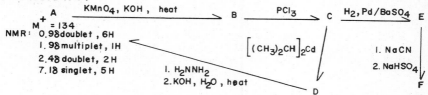

A
$M^+ = 134$
NMR: 0.98 doublet, 6H
1.98 multiplet, 1H
2.48 doublet, 2H
7.18 singlet, 5H

125

First Examination (D)   Answer Set

1. For inversion: thionyl chloride with pyridine

   R'R(H)C–OH + SOCl$_2$ $\longrightarrow$ R'R(H)C–O–S(=O)Cl (with pyridine, Cl$^-$, pyridinium) $\longrightarrow$ Cl–C(R')(R)(H) + SO$_2$ + pyridinium

   For retention: thionyl chloride with an inert solvent and heating

   R'R(H)C–OH + SOCl$_2$ $\xrightarrow{\text{heptane}}$ R'R(H)C–O–S(=O) · HCl $\longrightarrow$ R'R(H)C–Cl + SO$_2$

   In pyridine (solvent) the hydrogen chloride generated in the initial reaction forms pyridinium chloride. The chloride ion, maintained in the reaction system this way, is available to perform a displacement reaction on the ester. In the absence of pyridine the hydrogen chloride does not dissociate and the ester decomposes thermally through an intimate ion pair.

2. (a)

   (1) CH$_3$–O–OH + HOCH$_2$CH(OH)–

   (2) ester with β-OH

   (3) diester with OEt groups

   (4) diester with terminal OH

   (b)

   (1) isobutylene (CH$_3$)$_2$C=CH$_2$

   (2) neopentyl methyl ketone

   (3) neopentyl bromide

3. (a) t-Bu–O–cyclohexyl $\xleftarrow[\text{2. NaBH}_4]{\text{1. Hg(O}_2\text{CCF}_3)_2,\ (CH_3)_2C=CH_2}$ cyclohexanol $\xrightarrow{\text{NaBH}_4 \text{ or LiAlH}_4 \text{ or H}_2, \text{Pd/C}}$ cyclohexanone

   Formation of the tertiary ether is done in the shortest way by this method. An alternative route is to use the solvomercuration-demercuration with <u>t</u>-butyl alcohol and the alkene by dehydration of cyclohexanol.

(b)

[Scheme showing a diol with two CH₃ groups and two OH groups, converted via OsO₄ or KMnO₄, or via H₂O₂, HCO₂H, H₂O, from alkenes; alkene prepared by H₂, Pd/BaSO₄ or Li, liq NH₃ from an alkyne]

Either <u>trans-</u> or <u>cis-</u> hydroxylation reactions may be used depending on the geometry of the alkene used.

(c)

$HO_2C\!\!\smile\!\!\curlywedge \xleftarrow{H_3O^+} NC\!\!\smile\!\!\curlywedge \xleftarrow{NaCN,\,DMSO} Br\!\!\smile\!\!\curlywedge \xleftarrow{PBr_3} HO\!\!\smile\!\!\curlywedge$

A new carbon-carbon bond must be generated. It is accomplished by cyanide displacement of halogen. An alternative route is carbonation of the Grignard derived from the halide.

(d)

[Ar–CO–CH₂CH₂CH₃ ← (via R₂Cd) ← Ar–COCl ← SOCl₂ ← Ar–COOH ← KMnO₄ ← Ar–CH₂OH]

Again, a new carbon-carbon bond must be formed. For this an alkyl cadmium addition to an acyl halide is used. Grignard addition in this system would lead to the tertiary alcohol. An alternative route would involve initial oxidation to the aldehyde, Grignard addition, and reoxidation.

4.

A: isobutylbenzene
B: PhCOOH
C: PhCOCl
D: PhCOCH(CH₃)₂
E: PhCHO
F: PhCH(OH)COOH

For this problem the fundamental structural information is given with the species A in the form of an NMR and a molecular ion measurement from the mass spectrometer. The NMR clearly indicates the presence of a monosubstituted phenyl and an isobutyl function. From the molecular ion one notes that no other atoms can be present.

First Examination (E)   One Hour

For each of the questions 1-10 choose the set of reagents which best accomplishes the desired conversion.

1. 1-octene ⟶ 1-octanol
   (a) sulfuric acid, water
   (b) 1. hydrogen peroxide, water 2. potassium hydroxide
   (c) potassium permanganate, potassium hydroxide, water
   (d) 1. diborane 2. hydrogen peroxide, water, potassium hydroxide
   (e) 1. diborane 2. water

2. cyclohexanone ⟶ cyclohexane
   (a) 1. lithium aluminum hydride 2. potassium hydroxide, water
   (b) 1. hydrazine 2. potassium hydroxide, water
   (c) 1. carbon tetrachloride, triphenyl phosphine 2. hydrogen, platinum
   (d) hydrogen, palladium on calcium carbonate
   (e) sodium borohydride, potassium hydroxide

3. 1-bromoheptane ⟶ octanoic acid
   (a) 1. sodium cyanide 2. sulfuric acid, water
   (b) 1. magnesium, ether 2. formaldehyde
   (c) 1. methyl lithium 2. potassium permanganate
   (d) 1. lithium 2. formic acid
   (e) carbon dioxide, sulfuric acid

4. cyclohexanol ⟶ cyclohexyl 2-butyl ether
   (a) 1. 2-butanol, potassium hydroxide 2. lithium aluminum hydride
   (b) 1. sodium hydride 2. 2-bromobutane
   (c) 1. phosphorus tribromide 2. 2-butanol
   (d) 1. 2-butanol, mercuric trifluoroacetate 2. sodium borohydride
   (e) 1. 2-butene, mercuric trifluoroacetate 2. sodium borohydride

5. propionic acid ⟶ 3-pentanone
   (a) 1. acetic acid, phosphorus pentoxide 2. potassium permanganate
   (b) 1. thionyl chloride 2. diethyl cadmium
   (c) 1. thionyl chloride 2. ethanol 3. hydrogen peroxide
   (d) ethyl lithium
   (e) 1. thionyl chloride 2. ethyl magnesium bromide

6. 2,2-dimethyl-1-propanol ⟶ 2,2-dimethyl-1-bromopropane
   (a) carbon tetrabromide, triphenyl phosphine
   (b) phosphorus tribromide, potassium hydroxide
   (c) phosphorus tribromide, hν, bromine
   (d) carbon tetrabromide, hν
   (e) bromine

7. 2-phenylethanol ⟶ phenylacetaldehyde
   (a) potassium permanganate

(b) chromic anhydride, sulfuric acid
   (c) DCC, DMSO
   (d) osmium tetroxide
   (e) ozone, potassium hydroxide

8. 2-hexyne ⟶ racemic 3,4-dihydroxyhexane
   (a) 1. hydrogen, palladium on barium sulfate 2. osmium tetroxide
   (b) potassium permanganate, potassium hydroxide
   (c) 1. hydrogen, platinum 2. potassium permanganate
   (d) 1. lithium, liquid ammonia 2. osmium tetroxide
   (e) 1. lithium aluminum hydride 2. osmium tetroxide

9. 3-methyl-2-butanone ⟶ isobutyric acid
   (a) hydrogen peroxide, potassium hydroxide, water
   (b) 1. lithium aluminum hydride 2. potassium hydroxide
   (c) phosphorus tribromide, bromine
   (d) 1. hydrogen peroxide 2. sulfuric acid, water
   (e) bromine, potassium hydroxide

10. 3,3-dimethyl-1-butene ⟶ 3,3-dimethyl-1-butanol
    (a) 1. mercuric acetate, water 2. sodium borohydride
    (b) sulfuric acid, water
    (c) osmium tetroxide, potassium hydroxide
    (d) 1. magnesium, ether 2. water
    (e) 1. sodium acetate, potassium hydroxide 2. lithium aluminum hydride

For each of questions 11-20 choose the <u>major</u> organic product(s) in the reaction indicated.

11. benzyl phenyl ether + hydrogen bromide ⟶
    (a) benzyl bromide + bromobenzene
    (b) benzyl alcohol + bromobenzene
    (c) benzyl bromide + phenol
    (d) benzyl bromide + o-dibromobenzene
    (e) benzyl bromide + 1,2-dibromocyclohexane

12. toluene + 1. chlorine 2. water, potassium hydroxide ⟶
    (a) benzyl chloride
    (b) potassium benzoate
    (c) 1,2-diphenyl-1,2-dichloroethane
    (d) benzyl alcohol
    (e) hexachlorobenzene

13. propionic acid + ethyl magnesium bromide ⟶
    (a) propane + acetic acid
    (b) iodomagnesium propionate + ethane
    (c) 3-pentanone
    (d) pentanoic acid
    (e) pentane

14. benzaldehyde + formaldehyde + conc. sodium hydroxide ⟶
    (a) 1,2-dihydroxy-1-phenylethane
    (b) no reaction
    (c) styrene
    (d) benzoic acid
    (e) benzyl alcohol

15. toluene + 1. chromic anhydride, acetic anhydride 2. aq. acid ⟶
    (a) benzoic acid
    (b) benzaldehyde
    (c) acetic acid
    (d) acetaldehyde
    (e) 1-hydroxy-1-phenylethane

16. t-butyl chloride + sodium methoxide ⟶
    (a) 2-methylpropene
    (b) methyl t-butyl ether
    (c) 2-methylpropane
    (d) dimethyl ether
    (e) methyl chloride

17. meso-2,3-epoxybutane + potassium hydroxide, water ⟶
    (a) no reaction
    (b) butanedione
    (c) 1,3-butadiene
    (d) racemic-2,3-dihydroxybutane
    (e) meso-2,3-dihydroxybutane

18. isobutyric acid + phosphorus tribromide + bromine ⟶
    (a) isobutyl dibromo phosphine
    (b) 2-methylpropene
    (c) 2-bromo-2-methylpropionic acid
    (d) 2-methyl-1,2-dibromopropane
    (e) 2-methyl-1-bromopropane

19. potassium acetate + potassium ethoxide ⟶
    (a) ethyl acetate
    (b) acetic acid
    (c) ethyl potassium
    (d) ethanol
    (e) no reaction

20. 2-pentanol + iodine, potassium hydroxide ⟶
    (a) 2-iodopentane
    (b) 3-pentanol
    (c) potassium butanoate
    (d) pentanoic acid
    (e) 1-iodopentane

For each of questions 21-25 choose the reagent which will serve to differentiate most quickly and clearly the indicated compounds.

21. 2-butanol and 3-pentanol
    (a) sulfuric acid, chromic anhydride
    (b) bromine, carbon tetrachloride
    (c) hydrogen bromide, acetic acid
    (d) iodine, potassium hydroxide
    (e) ethanol, hydrogen chloride

22. t-butyl alcohol and 1-butanol
    (a) dry hydrogen bromide
    (b) ozone, acetic acid
    (c) thionyl chloride
    (d) zinc chloride, hydrogen chloride
    (e) bromine, carbon tetrachloride

23. 2-butanol and 2,3-dihydroxybutane
    (a) sodium
    (b) iodine, potassium hydroxide
    (c) periodic acid
    (d) zinc chloride, hydrogen chloride
    (e) acetone, zinc chloride

24. benzoic acid and acrylic acid
    (a) p-hydroxyaniline
    (b) thionyl chloride
    (c) bromine, carbon tetrachloride
    (d) zinc chloride, hydrogen chloride
    (e) potassium hydroxide, water

25. anisole and ethyl benzene
    (a) hydrogen chloride
    (b) sulfuric acid
    (c) aluminum chloride, ethyl iodide
    (d) bromine, carbon tetrachloride
    (e) bromine, hν

First Examination (E)   Answer Set

1.(d) 2.(b) 3.(a) 4.(e) 5.(b) 6.(a) 7.(c) 8.(d) 9.(e) 10.(a) 11.(c) 12.(b)
13.(b) 14.(a) 15.(b) 16.(a) 17.(d) 18.(c) 19.(e) 20.(c) 21.(d) 22.(d) 23.(c)
24.(c) 25.(b)

Second Examination (A) One Hour

1. (25 pts, 14 min) Predict the major organic product in each of the following reactions:
   (a) CH_3OCH_2CN $\xrightarrow{\text{1. PhMgBr} \quad \text{2. H}_2\text{O}}$

   (b) R-2-methylbutamide $\xrightarrow{\text{NaOH, H}_2\text{O, Br}_2}$

   (c) Cyclohexanone $\xrightarrow{(Et)_2NH_2Cl,\ H_2CO}$

   (d) 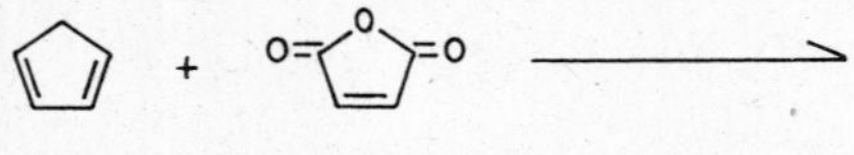 (cyclopentadiene) + (maleic anhydride) $\longrightarrow$

2. (20 pts, 12 min) Explain briefly the following chemical observations. Properly annotated diagrams or structures should be used.
   (a) p-Nitrophenol is a stronger acid than phenol.
   (b) Acetylacetone, as a pure liquid, exists principally in the enol form.
   (c) Triphenylamine is a much weaker base than triethylamine.
   (d) Phenol does not react with benzenediazonium chloride when added to it at pH=11.

3. (30 pts, 20 min) Give reasonable syntheses for each of the following compounds from the indicated starting materials and any other reagents thought necessary.
   (a) $\sim\sim NH_2$ from $\sim\sim Br$
   (b) PhCH=CHCH_3 (styrene-methyl) from PhCH(NH_2)CH_3
   (c) $\text{(isopropyl)}CH_2COOH$ from $\text{(isopropyl)}Br$
   (d) $\text{(t-butyl)}COOH$ from $\text{(t-butyl)}CHO$
   (e) (3-ethylpentan-... diketone structure) from (methyl acetoacetate)

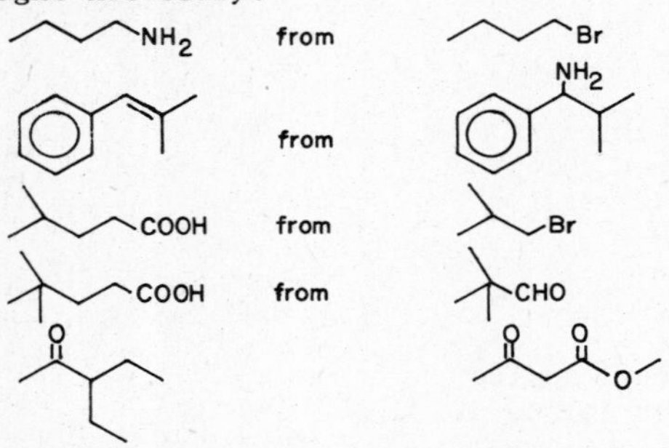

4. (25 pts, 14 min) Give the correct structures for compounds A-F.

   (A) C_7H_6O $\xrightarrow{\text{1. BrCH}_2\text{CO}_2\text{Et, Zn} \quad \text{2. H}_2\text{O}}$ (B) $\xrightarrow{\text{MnO}_2, \text{pentane}}$ (C)

   NMR 7.2δ, 5H, singlet
   10.0δ, 1H, singlet

   (B) IR(partial) 1735 cm^{-1}, 3350 cm^{-1}

   (C) IR(partial) 1735 cm^{-1}, 1750 cm^{-1}, $M^+ = 192$

   $\xrightarrow{H_3O^+}$

   $\xrightarrow{Ph_3P\,CH_2}$ (F)

   (E) $\xleftarrow{\text{KOH, "A"}}$ (D)

   $M^+ = 208$
   $\lambda_{max} = 225$ nm

   (D) NMR 2.2δ, 3H, singlet
   7.1δ, 5H, singlet

   (F) IR(partial) 1735 cm^{-1}, 1650 cm^{-1}

Second Examination (A)   Answer Set

1. (a) CH₃OCH₂C(=O)-Ph ⟵ 1. PhMgBr 2. H₂O ⟵ CH₃O CH₂CN

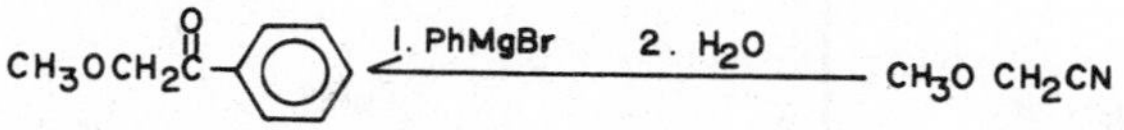

The important point to be recognized here is that in contrast to the reaction of Grignards with other derivatives of carboxylic acids, the initial reaction will produce a species that is not subject to further reaction with a carbanion-like reagent. This species (shown below) will first accept a proton from water and then be hydrolyzed.

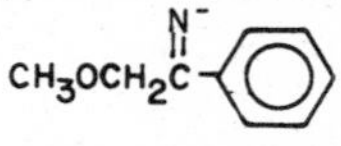

CH₃OCH₂C(=N⁻)-Ph

(b) CH₃CH₂-C(CH₃)H-NH₂ ⟵ H₂O, NaOH, Br₂ ⟵ CH₃CH₂-C(CH₃)H-CONH₂

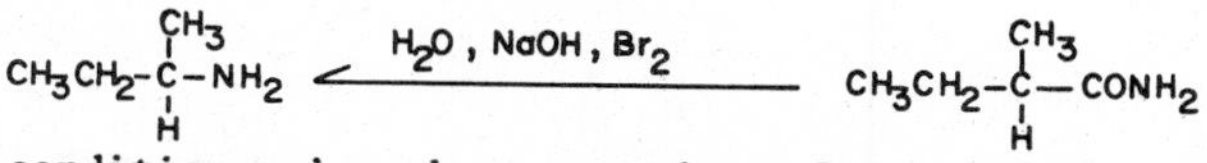

The conditions given here are those for the Hofmann degradation of amides. It should be noted that this reaction proceeds with migration of the group originally attached to the carboxyl carbon from that site to nitrogen with retention of configuration about the migrating site. The resultant isocyanate is then hydrolyzed to the free amine without disturbing the chiral center.

(c) [cyclohexanone with CH₂-N(Et)₂ substituent] ⟵ (Et)₂NH₂Cl, H₂CO ⟵ [cyclohexanone]

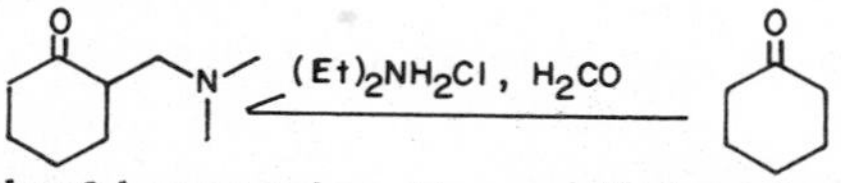

One should recognize the conditions here as those for a Mannich reaction, the overall course of which may be represented as

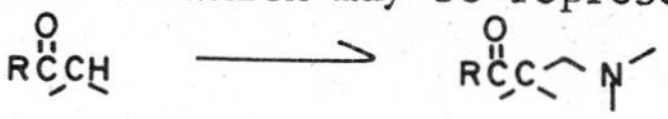

RCCH₃ (O) ⟶ RCC(=O)CH₂~N

(d) [endo bicyclic anhydride product] ⟵ [cyclopentadiene] + [maleic anhydride, O=C-O-C=O]

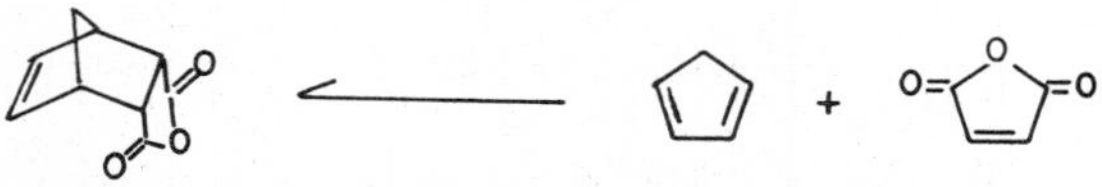

A Diels-Alder reaction is involved here, the cyclopentadiene adding to the dienophile, maleic anhydride. The stereochemical aspects are quite important; the product has the endo configuration, the carbon-containing substituent of the bicyclic system being on the "inner" or more hindered side. This may be seen in the following view where the carbon-containing substituent is on the "two-carbon side" of the "bridge".

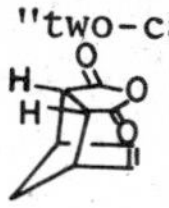

[endo structure diagram]

2. (a) The p-nitrophenol is a stronger acid than phenol owing to stabilization (by resonance, as shown below) of the product anion involving the nitro group.

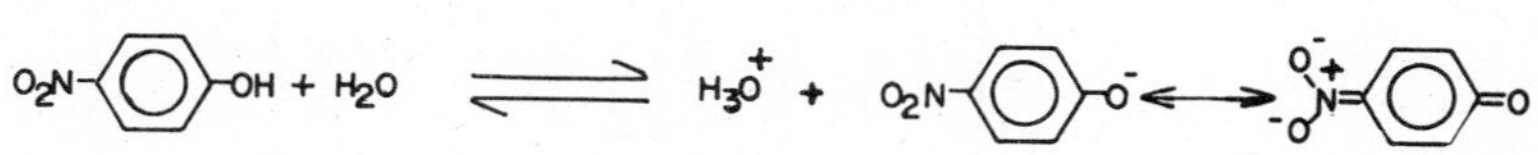

O₂N-C₆H₄-OH + H₂O ⇌ H₃O⁺ + O₂N-C₆H₄-O⁻ ⟷ ⁻O-N⁺(=O)=C₆H₄=O

133

(b) There is stabilization of the enol form of acetylacetone (shown below) by the intramolecular hydrogen bonding and conjugation of the olefinic linkage with the remaining carbonyl.

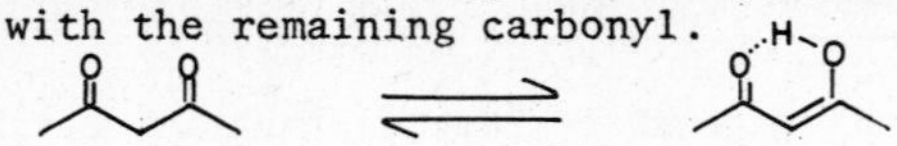

(c) In the case of triphenylamine there is very significant electron withdrawl (by resonance, as shown below) from the nitrogen by the aromatic rings rendering the nitrogen less "electron rich" than it is in the triethylamine.

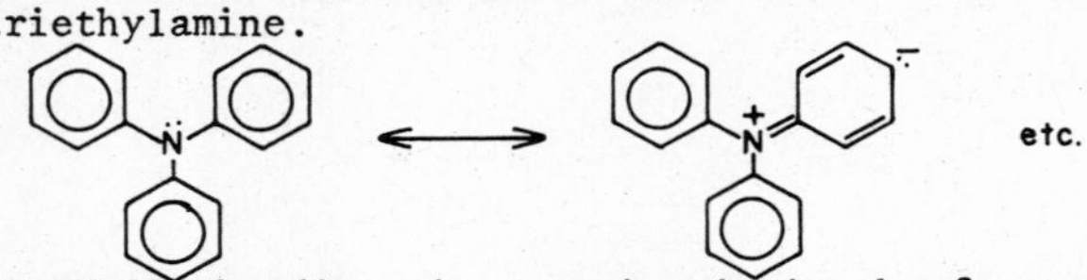

etc.

(d) At pH=11 the diazonium species is in the form of the hydroxide which is not able to undergo the coupling reaction.

3. (a)

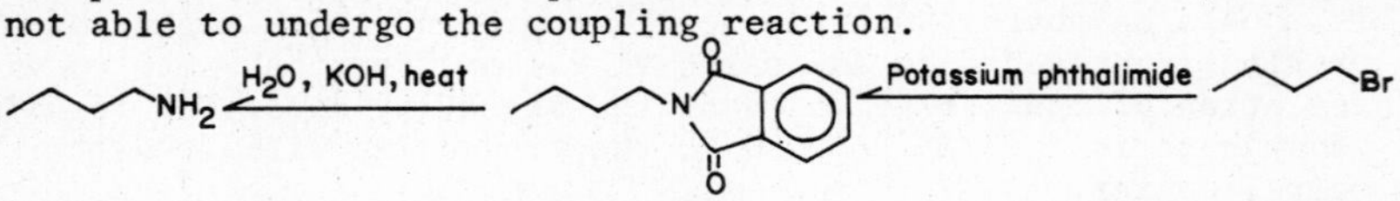

The Gabriel synthesis is the preferred route, as shown here. Attempting to obtain the mono-alkylated species by using ammonia, even in a large excess would give poorer results.

(b)

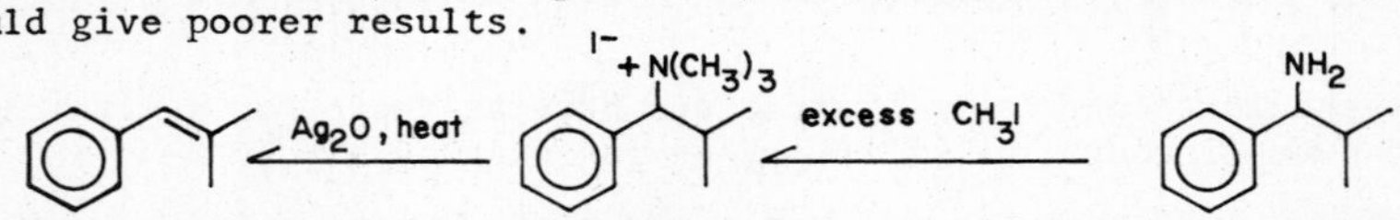

An elimination reaction is obviously in order here to generate the olefin linkage. In order to accomplish this the amine function must first be quaternized so that the nitrogen may leave as a neutral species. In the second step the quaternary ammonium hydroxide is formed and thermally decomposed.

(c)

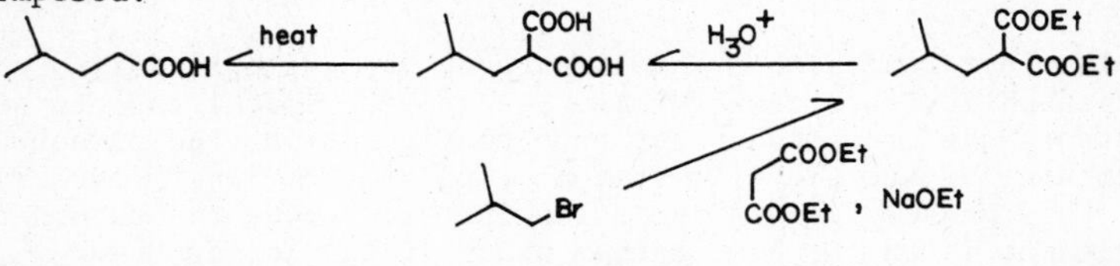

This is a classic example of the use of the malonic ester synthesis. The carbon chain is extended by two with a carboxylate terminus.

(d)

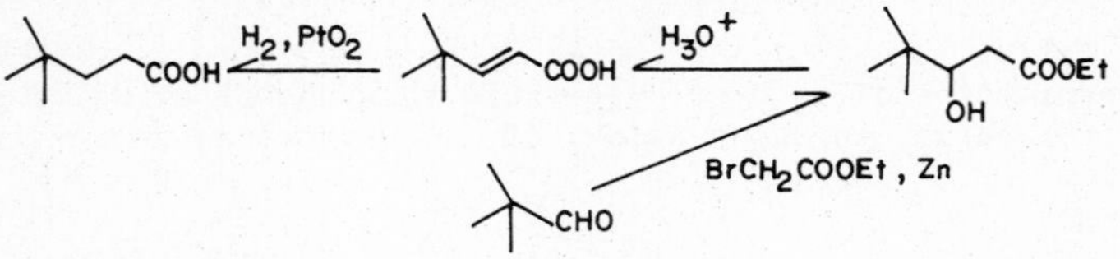

A Reformatsky reaction is used here to generate the fundamental carbon structure. Care must be taken in the method chosen to form the new carbon-carbon bond. Crossed Claisen reactions do not work well with aldehydes.

(e) [structure] $\xleftarrow{H_3O^+, heat}$ [structure with COOEt] $\xleftarrow{NaH, EtI}$ [structure with COOEt]

$\xleftarrow{NaH, EtI}$ [structure with COOEt]

This synthesis involves a minor extension of the standard acetoacetic ester route. Two successive substitutions are required prior to decarboxylation.

4. [structures A (benzaldehyde-CHO), B (PhCH(OH)CH2COOEt), C (PhCOCH2COOEt), D (PhCOCH3), E (PhCOCH=CHPh), F (PhC(=CH2)CH2COOEt)]

Using the approach of starting at the point where the greatest amount of structural information is given, compound A would be the first structure deduced. A complete nmr is given for A, indicating that only aromatic and aldehydic protons are present. Given the formula, compound A could only be benzaldehyde. Once this is determined the other structures follow reasonably. B is formed by a Reformatsky reaction leading to the secondary alcohol (benzylic). This is oxidized under extremely mild conditions to the β-ketoester C; under these conditions no cleavage of C-C bonds or ester hydrolysis would occur. D is formed by hydrolysis and subsequent decarboxylation of the β-ketoester. Compound E is formed by a crossed aldol condensation accompanied by dehydration, as is common in this type of reaction. Finally, F is formed by a Wittig reaction on C; the phosphorus ylid reacts with the ordinary carbonyl, but not with the ester linkage.

Second Examination (B) One Hour

1. (20 pts, 12 min) Predict the major organic product for each of the following reactions. In each case draw its structure.
   (a) C_6H_5MgBr 1. C_6H_5CN 2. H_3O^+
   (b) furfural $\xrightarrow{Ac_2O,\ NaOAc}$
   (c) $(CH_3)_3\overset{O}{C}CCH_3$ $\xrightarrow{H_2CO,\ (CH_3)_2NH_2Cl}$
   (d) 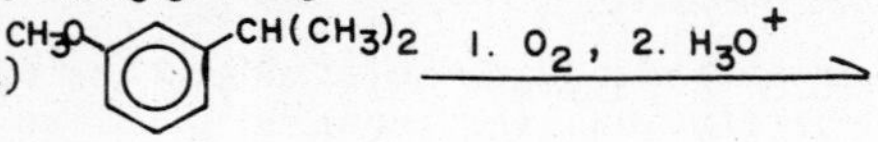 3-methoxy-1-isopropylbenzene 1. O_2, 2. H_3O^+

2. (25 pts, 12 min) Aniline, after treatment with sodium nitrite and sulfuric acid at 0°, reacts with phenol to yield the azo compound illustrated. Show all steps in the mechanism of its formation from the starting materials.

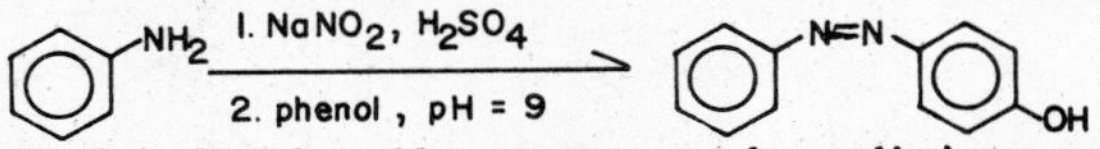

   $C_6H_5NH_2 \xrightarrow[\text{2. phenol, pH = 9}]{\text{1. NaNO}_2,\ H_2SO_4} C_6H_5-N=N-C_6H_4-OH$

3. (25 pts, 15 min) Supply all reagents and conditions necessary for the conversions I-V. More than one step may be required for each conversion. Where several routes are available, use the best.

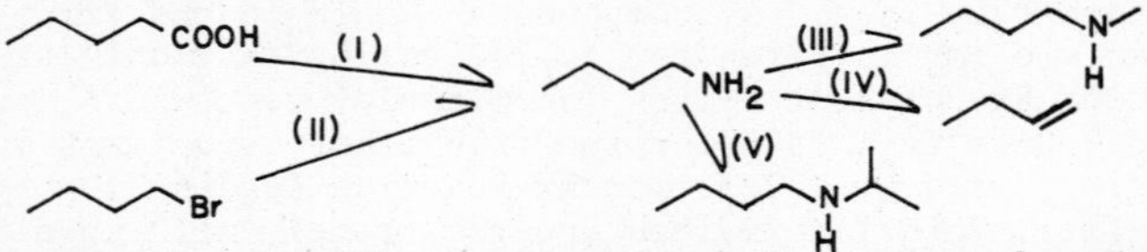

4. (30 pts, 18 min) Give reagents and conditions for the synthesis of each of the following compounds from the indicated starting materials, using any other reagents thought necessary. Several steps may be required for each synthesis.
   (a) 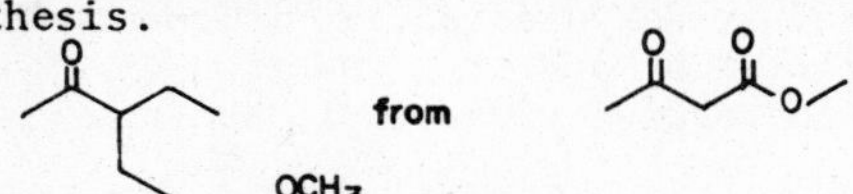 2-methylpentan-3-one (or similar ketone) from methyl acetoacetate
   (b) 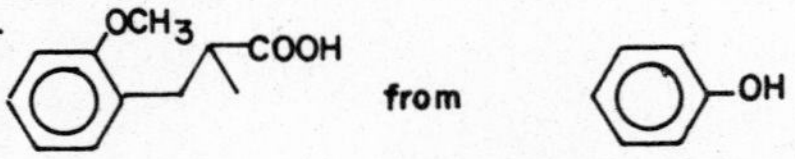 (2-methoxyphenyl)acetic acid from anisole; and phenol from benzene derivative shown
   (c) 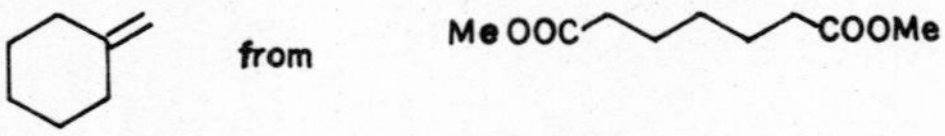 methylenecyclohexane from dimethyl glutarate (MeOOC-(CH_2)_3-COOMe)
   (d) 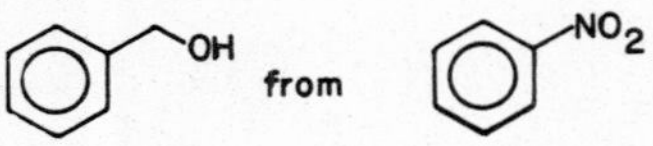 benzyl alcohol from nitrobenzene

Second Examination (B)   Answer Set

1. (a) [Ph-CO-Ph (benzophenone)]

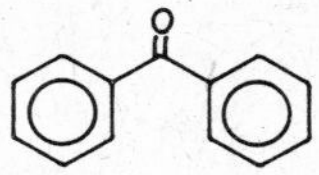

   As discussed previously it is important to note that the reaction of an
   organometallic with a nitrile will stop after mono-addition, that is,
   with the following species: [Ph-C(=N⁻MgBr⁺)-Ph]

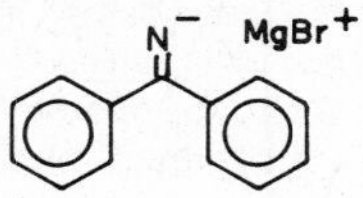

   which is later hydrolyzed.

   (b) [2-furyl-CH=CH-COOH]

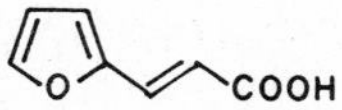

   This is an example of the Perkin condensation, one of the many modifications of the aldol condensation. It involves an aldehyde having no
   α-hydrogens in reaction with an anhydride (which is the enolate anion
   source) and a base, the carboxylate anion derived from the acid related
   to the anhydride.

   (c) [(CH₃)₃C-CO-CH₂-CH₂-NR₂]

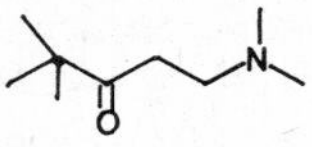

   As mentioned previously, this represents a Mannich reaction wherein the
   function $-CH_2NR_2$ is substituted for an active α-hydrogen of a ketone.

   (d) [3-methoxyphenol: CH₃O-C₆H₄-OH]

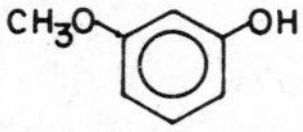

   This is another example of the conversion of a substituted cumene to a
   phenol. An additional problem for the student to consider here is the
   synthesis of the starting material. This is by no means as simple as the
   examination question.

2. In acid solution the nitrite ion is protonated to HONO. From this point
   the sequence is as follows:

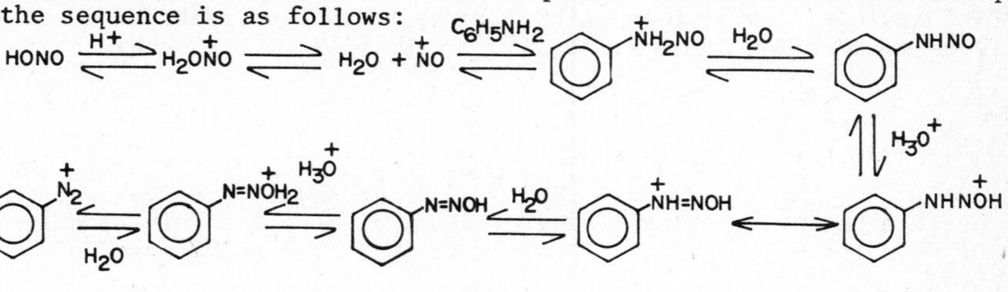

   $HONO \xrightleftharpoons{H^+} H_2ONO^+ \rightleftharpoons H_2O + NO^+ \xrightarrow{C_6H_5NH_2} C_6H_5\overset{+}{N}H_2NO \xrightarrow{H_2O} C_6H_5NHNO$

   $\Updownarrow H_3O^+$

   $C_6H_5\overset{+}{N}_2 \rightleftharpoons[H_2O]{} C_6H_5N=\overset{+}{N}OH_2 \xrightarrow{H_3O^+} C_6H_5N=NOH \xrightarrow{H_2O} C_6H_5\overset{+}{N}H=NOH \leftrightarrow C_6H_5\overset{+}{N}HNOH$

137

[Ph-N₂⁺] + [Ph-O⁻] ⟶ [Ph-N=N-C₆H₄=O (H)] ⟶ [Ph-N=N-C₆H₄-OH]

This is another example of a complex reaction that is explained by simple acid-base reactions. The critical aspects are realizing that the benzenediazonium ion must be formed from aniline for the ultimate coupling, and that the nitrosonium ion is generated intially. This species then reacts with the electron pair on the nitrogen atom of aniline yielding a species containing the nitrogen-nitrogen linkage. This then must lose the elements of water in several steps. Then occurs electrophilic substitution by the so-formed benzenediazonium ion on the phenoxide ion. One should note that reaction occurs on the phenoxide ion rather than on phenol itself.

3. I. R-NH₂ $\xrightarrow{\text{NaOH, H}_2\text{O}}_{\text{Br}_2}$ R-CONH₂ $\xrightarrow[\text{2. NH}_3]{\text{1. PCl}_3}$ R-COOH

First one should notice that the target amine bears one carbon less than the starting acid. The required reaction involves cleavage of a carbon-carbon bond which is accomplished by the Hofmann rearrangement. This rearrangement is set up by the preparation of the amide from the acid.

II. R-NH₂ $\xleftarrow{\text{H}_2\text{NNH}_2}$ R-N(phthalimide) $\xleftarrow{\text{Potassium phthalimide}}$ R-Br

This is simply the generation of a primary amine from a primary alkyl halide. The Gabriel synthesis is best for this. The use of ammonia is troublesome, as higher alkylated amines may be formed. Other cleavage methods may be used in the last step.

III. R-NHR' $\xleftarrow[\text{1. H}_2\text{CO, 2. HCOOH}]{}$ R-NH₂

The indicated conditions of the Clarke-Eschweiler reaction are better for <u>mono</u>-methylation of an amine than by reaction with a methyl halide.

IV. R-CH=CH₂ $\xleftarrow{\text{Ag}_2\text{O, heat}}$ R-N⁺(CH₃)₃ I⁻ $\xleftarrow{\text{excess CH}_3\text{I}}$ R-NH₂

An elimination reaction involving nitrogen requires initial quaternization of nitrogen. Elimination in this case is by the Hofmann method.

V. [scheme: CH₃CH₂CH₂-N(H)-CH(CH₃)₂ →(H₂, Pd)← CH₃CH₂CH₂-N=C(CH₃)₂ →(acetone, TsOH)← CH₃CH₂CH₂-NH₂]

Again, mono-alkylation of an amine is required, and again, a method not permitting additional alkylation is preferred. This is given by imine formation followed by reduction.

4. (a) [scheme: ethyl-substituted β-ketoester ←(H₃O⁺, heat)— diethyl-substituted malonate-type diester ←(1. NaH, 2. EtI, 3. NaH, 4. EtI)— acetoacetic ester]

This synthesis involves a minor extension of the standard acetoacetic ester route. Two successive substitutions are required prior to decarboxylation.

(b) [scheme: o-methoxy arylpropanoic acid →(1. H₂, Pd; 2. H₃O⁺)← o-methoxycinnamate (H₃CO-C₆H₄-CH=CH-COOMe) ←(1. CH₃CHBrCOO-; 2. Zn, then H₃O⁺)— o-methoxy-benzaldehyde (OCH₃, CHO) →(CH₃I, NaOH)← o-hydroxybenzaldehyde (OH, CHO) ←(NaOH, CHCl₃, H₂O)— phenol]

An overall view of this system, target molecule and starting material, indicates that an *ortho*-specific substitution on phenol is required, such as the Reimer-Tiemann process, which places an aldehyde function on the ring. This is then functionalized using a Reformatsky reaction.

(c) [scheme: methylenecyclohexane ←(Ph₃PCH₂)— cyclohexanone ←(H₃O⁺, heat)— 2-carbomethoxycyclohexanone ←(NaOMe)— MeO₂C(CH₂)₅CO₂Me]

An overview of this system indicates that some cyclization process is required. The carbon number is appropriate for the Dieckmann condensation, a six-membered ring formed from a seven-carbon diacid derivative. Once the ring is formed a decarboxylation is performed and the carbonyl group is converted to the alkene by a Wittig reaction.

(d) [scheme: PhOH ←(LiAlH₄)— PhCOOH ←(H₃O⁺)— PhCN ←(1. NaNO₂, H₂SO₄; 2. Cu₂(CN)₂)— PhNH₂ ←(Sn, HCl)— PhNO₂]

Substitution of a nitrogen function on the aromatic ring is required with the formation of a carbon-carbon bond. Several routes are possible once the diazonium ion is generated. The one shown requires the fewest steps. A Sandmeyer reaction is performed, followed by hydrolysis and reduction.

Second Examination (C)   One Hour

1. (25 pts, 15 min) A second-semester organic chemistry student (who passed
   the first semester with a C-) was assigned the task of synthesizing
   p-bromo-(2-propenyl)benzene starting with aniline. He chose the following
   route:

   What is wrong with this route? Suggest one which would accomplish the
   desired result.

2. (25 pts, 15 min) Supply the reagents and conditions for conversions I-V.
   Note that more than one step may be needed for each conversion.

3. (25 pts, 15 min) Devise syntheses of (a) p-methylbenzyl amine, (b)
   m-hydroxystyrene, and (c) methyl cyclopentyl ketone, starting with
   benzene, toluene, aliphatic compounds of five carbons or less, and any
   inorganic reagents thought necessary.

4. (25 pts, 15 min) Give the correct structures for compounds A-G.

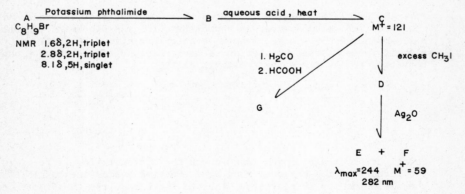

Second Examination (C)   Answer Set

1. Two major errors should be evident immediately upon inspection of the
    proposed route. First, bromination of aniline directly will lead to
    2,4,6-tribromoaniline. Even if one tries to control the reaction by
    limiting the amount of bromine, a horrendous mixture will result. The
    nitrogen should first be acylated, somewhat deactivating the ring, before
    bromination, and then be deacylated. Second, the reaction of the Grignard
    with the acid will result simply in formation of the carboxylate ion and
    methane. Once the carboxylate ion is formed no attack by the Grignard
    will occur. This problem may be circumvented either by formation of the
    ester prior to reaction with the Grignard, or by the use of an alkyl-
    lithium reagent, which is reactive enough to attack a carboxylate anion,
    although one molar equivalent is "wasted" in reaction with the acidic
    proton. An acceptable route would be

    [Scheme: PhNH₂ →(Ac₂O)→ PhNHAc →(Br₂)→ 4-Br-C₆H₄-NHAc →(H₃O⁺)→ 4-Br-C₆H₄-NH₂ →(1. NaNO₂, H₂SO₄; 2. Cu₂(CN)₂)→ 4-Br-C₆H₄-CN →(H₃O⁺)→ 4-Br-C₆H₄-COOH]

2. I. [Scheme: PhCOCH₂COOEt ←(NaOEt)— CH₃COOEt + PhCOOEt]

    This is a typical crossed Claisen condensation. One should note that the
    base used corresponds to the alcohol portion of product and starting
    esters. If it doesn't, one may obtain mixtures of products that differ in
    their ester linkage. If a later hydrolysis were planned (as occurs in this
    particular problem) the formation of a mixture of esters would not present
    too severe a difficulty, except in the purification and identification
    of the intermediates.

    II. [Scheme: PhCONH₂ ←(NH₃)— PhCOCl ←(PCl₅)— PhCOOH ←(H₃O⁺)— PhCOOEt]

    A straightforward conversion of one carboxylate derivative to another is
    required here. This is accomplished by conversion of one (the ester) to
    the acyl chloride from which the product is obtained.

III. [structure: acetophenone] ⇌ (H₃O⁺, heat) [structure: ethyl benzoylacetate]

Here a β-ketoester is hydrolyzed and decarboxylated by heating.

IV. [structure: β-methylstyrene derivative] ⇌ Ph₃PC(CH₃)₂ [structure: acetophenone]

Again, a rather simple problem; it requires the recognition of the "carbonyl replacement" aspect for the Wittig reaction.

V. [structure: β-amino ketone] ⇌ (CH₃)₂NH₂Cl / H₂CO [structure: acetophenone]

If one has worked through the previous examinations, one should have recognized immediately that this is a Mannich condensation involving substitution of CH₂NR₂ for H at the α-position of a ketone.

3. (a) [p-tolyl-CH₂NH₂] ⇌ LiAlH₄ [p-tolyl-CN] ⇌ 1.NaNO₂, H₂SO₄ / 2.Cu₂(CN)₂ [p-toluidine] ⇌ 1.H₂SO₄,HNO₃ / 2.Sn, HCl [toluene]

This synthesis involves the generation of a carbon-carbon bond and the introduction of a nitrogen in the same step, that is, the Sandmeyer reaction. Conversion of the nitrile to the amine is accomplished by direct reduction.

(b) [m-vinylphenol] ⇌ 1.NaNO₂, H₂SO₄ / 2. H₂O, heat [m-vinylaniline] ⇌ Sn, HCl [m-vinylnitrobenzene]

[phenol] → Ac₂O / AlCl₃ → [m-hydroxyacetophenone] ⇌ 1.LiAlH₄ / 2.H₃O⁺ ← H₂SO₄, HNO₃ [nitro ketone] → 1.NaBH₄ / 2.H₃O⁺ → [product]

This is a complicated route requiring a proper understanding of the order in which the various reactions are performed. Acylation <u>precedes</u> nitration. The nitrogen is introduced so that it might be converted to a hydroxyl at the end. One must be careful of the conditions used to reduce the two groups, and depending on which are chosen, one must use a particular order. For this last problem two approaches are shown.

(c) [cyclopentyl methyl ketone] ⇌ CH₃COCl [(cyclopentyl)₂Cd] ⇌ CdCl₂ [cyclopentyl-MgBr] ⇌ Mg, Et₂O [cyclopentyl bromide]

This problem is quite simple compared to the others in this series. All that is required is recognition of a reaction that will allow generation of an aliphatic ketone.

4. (A) PhCH$_2$CH$_2$Br

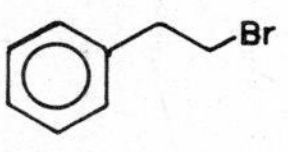

(B) PhCH$_2$CH$_2$-N(phthalimide)

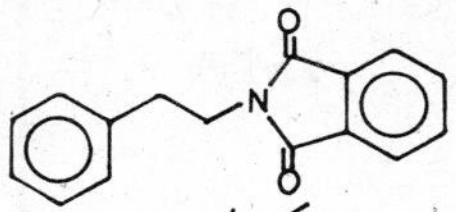

(C) PhCH$_2$CH$_2$NH$_2$

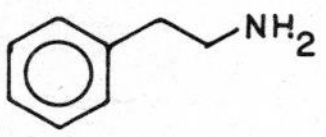

(D) PhCH$_2$CH$_2$-N$^+$(CH$_3$)$_2$ I$^-$ (with additional methyl)

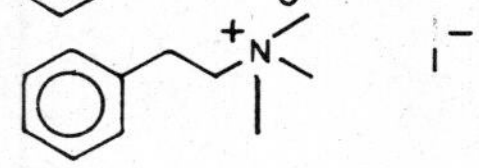

(E) PhCH=CH$_2$

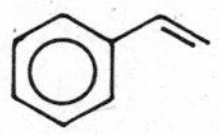

(F) N(CH$_3$)$_3$

(G) PhCH$_2$CH$_2$-NH-CH$_3$

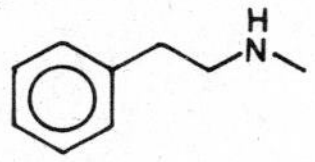

For this problem the greatest amount of structural information is given with compound A. A complete nmr spectrum is available which may be interpreted to indicate a mono-substituted benzene with a two-carbon substituent. That is the β-bromo structure as shown and not an α-bromo compound as may be deduced both the intensities and splittings of the two non-aromatic nmr signals. Compounds B and C are intermediate and product respectively in the Gabriel synthesis of amines. Compound G is the result of the Clarke-Eschweiler methylation, while D is the intermediate and E and F are the products of the Hofmann elimination.

Second Examination (D)   One Hour

1. (25 pts, 15 min) Give the correct structure for the organic product(s) of each of the following reactions.

   (a) cyclohexyl-CH₂-C(=O)-O-CH₂-? + furfural (furan-CHO) $\xrightarrow{NaOCH_3}$

   (b) cyclohexyl-NHCH₃ $\xrightarrow{1. H_2CO, HCOOH, 2. H_2O_2, 3. heat}$

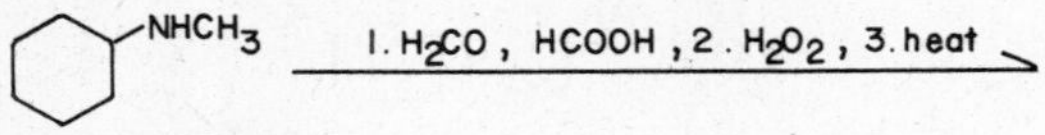

   (c) phenol $\xrightarrow{Br_2}$

   (d) benzaldehyde + methyl ethyl ketone $\xrightarrow{1. \text{piperidine, acetic acid} \quad 2. H_3O^+}$

   (e) 3-pentanone $\xrightarrow{1. \text{excess piperidine} \quad 2. \text{ethyl iodide} \quad 3. H_3O^+}$

2. (25 pts, 15 min) In a Perkin condensation cinnamic acid is obtained from benzaldehyde, acetic anhydride, and anhydrous potassium acetate. Give the detailed mechanism for this preparation.

3. (25 pts, 15 min) Give the correct structures for each of the compounds indicated A–H.

   A $\xrightarrow{P_2O_5, \text{ heat}}$ B $\xrightarrow{1. PhMgBr \quad 2. H_2O}$ C $\xrightarrow{Ph_3PCH_2, \text{ heat}}$ D
   C_4H_9NO C_4H_7N $C_{10}H_{12}O$

   NMR 1.2δ, 6H, doublet
       2.4δ, 1H, septet
       5.9δ, 2H, broad singlet

   C $\xrightarrow{H_2NOH}$

   G + H $\xrightarrow{1. H_3PO_4, \text{heat} \quad 2. \text{aq. acid}}$ E + F (isomers)
   C_3H_9N $C_7H_6O_2$

   NMR 0.7δ, 6H, doublet
       2.8δ, 1H, septet
       5.2δ, 2H, broad singlet

4. (25 pts, 15 min) Give syntheses for three of the following compounds from the indicated starting materials using any other reagents thought necessary. Note: Several steps may be required for each, and all reagents and conditions must be shown.

   (a) 4-chlorophenol (Cl, OH on benzene) from aniline ($C_6H_5NH_2$)

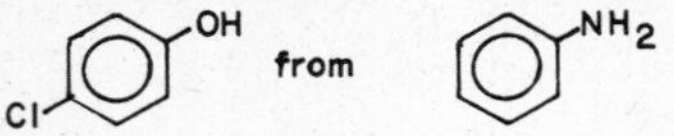

   (b) fluorobenzene from benzene

   (c) cyclohexylidene (methylenecyclohexane) from $MeO_2C(CH_2)_5CO_2Me$

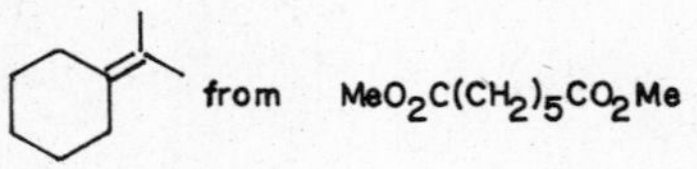

   (d) N-benzylaniline (PhCH₂NHPh) from benzyl bromide (PhCH₂Br)

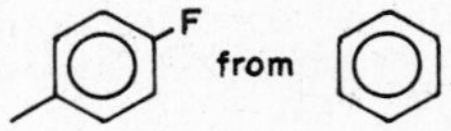

Second Examination (D)   Answer Set

1. (a) [structures: phenyl-furyl enol acetate + phenyl-furyl acrylate]

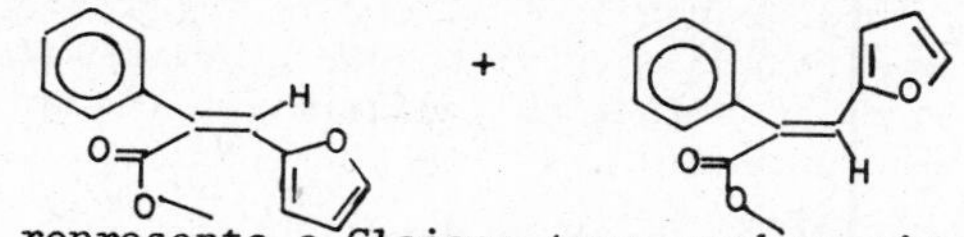

This represents a Claisen-type condensation with the phenylacetate serving as the carbanion precursor. Note the possibility of a pair of geometrical isomers being generated upon olefin formation.

(b) [cyclohexene] + Me_2NOH

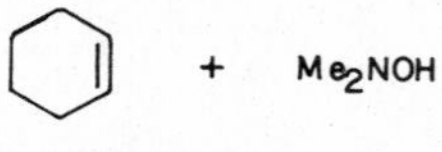

This system involves initial methylation of the amine (by the Clarke-Eschweiler route) followed by a Cope elimination proceeding through the amine oxide. As there are no β-hydrogens on the methyl groups, elimination can occur only on the cyclohexyl ring.

(c) [2,4,6-tribromophenol structure with Br, OH, Br, Br]

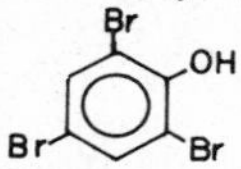

The hydroxyl on the phenyl ring so activates it for electrophilic aromatic substitution that the reaction proceeds to the tribromide.

(d) [PhCH=CH-C(=O)-Et + Ph-CH=CH-O-C(=O)-Me structures]

The conditions given are relatively mild for condensations. The reaction (Knoevenagel reaction) proceeds by initial condensation of the piperidine with the ketone.

(e) [3-methyl-2-pentanone structure]

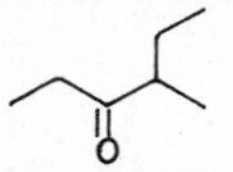

This reaction proceeds by generation of an enamine derived from the 3-pentanone and piperidine.

2. [mechanism sequence: acetic anhydride → NaOAc → carbanion ↔ enolate → PhCHO → aldol adduct → ... → cinnamic acid (COOH) via intermediates with OAc, HB]

145

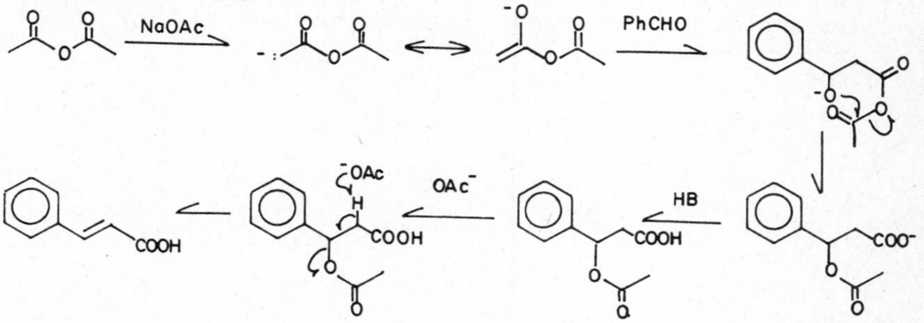

For this condensation the acetic anhydride serves as the carbanion precursor and the acetate ion as the base. The carbanion, once generated, reacts with the available aldehyde to give a more or less normal aldol-type adduct. However, there is an intramolecular ester formation due to the alkoxide site being in close proximity to the remaining anhydride linkage. Elimination occurs last.

3.

A: (CH₃)₂CH-C(=O)-NH₂
B: (CH₃)₂CH-CN
C: (CH₃)₂CH-C(=O)-O-Ph
D: (CH₃)₂C=C(-O-Ph)... (isopropenyl phenyl ester)
E: (CH₃)₂CH-C(=N-OH)-Ph
F: (CH₃)₂CH-C(=N-OH)-Ph (other oxime isomer, HO-N)
G: (CH₃)₂CH-NH₂
H: Ph-COOH

One looks for critical structural data here and finds nmr information for two species. Both A and G exhibit characteristics of an isopropyl group. As G necessarily is isopropyl amine, with the added elemental composition, A must be the amide of isobutyric acid. With this data the remainder is quite clear until one gets to the point of deciding which of the oxime isomers is E and which is F. This is decided on the basis that in generating G and H the group anti- to the hydroxyl migrates.

4. (a) PhNH₂ →[Ac₂O] PhNHC(=O)CH₃ →[Cl₂] 4-Cl-C₆H₄-NHC(=O)CH₃ →[H₃O⁺] 4-Cl-C₆H₄-NH₂

→[1. NaNO₂, H₂SO₄; 2. H₂O, heat] 4-Cl-C₆H₄-OH

The approach here is to start at the target molecule and work to the starting material. The hydroxyl is best introduced last as a replacement of the amino function. Prior to this care must be exercised in the introduction of the halogen, the amino group being deactivated.

(b) C₆H₆ →[CH₃I / AlCl₃] C₆H₅CH₃ →[H₂SO₄ / HNO₃] 4-O₂N-C₆H₄-CH₃ →[Sn, HCl] 4-H₂N-C₆H₄-CH₃

→[1. H₂SO₄, NaNO₂; 2. HBF₄; 3. heat] 4-F-C₆H₄-CH₃

The introduction of the halogen is most reasonably accomplished by diazotization of an amino function. Starting at the target molecule one readily sees that this is best done as the final step in the sequence.

For the remainder, the methyl group should be introduced <u>before</u> the nitro as this obviates problems of reactivity and orientation.

(c) [Diethyl pimelate] →(NaOMe)→ [2-carbomethoxycyclohexanone] →(H$_3$O$^+$, heat)→ [cyclohexanone] →(Ph$_3$P C(CH$_3$)$_2$)→ [isopropylidenecyclohexane]

The Dieckmann condensation should immediately come to mind upon seeing a cyclic system being produced from an α,ω-dicarboxylic acid. The remainder of the problem is the proper functionalization of the ring. Cyclohexanone is generated by deesterification and decarboxylation, and the final product is formed in a Wittig reaction.

(d) [Benzyl bromide] →(Potassium phthalimide)→ [N-benzylphthalimide] →(KOH, H$_2$O)→ [benzylamine, PhCH$_2$NH$_2$]

→(1. H$_2$CO, 2. HCOOH)→ [PhCH$_2$NHCH$_3$]

Here one must avoid using simple "alkyl halide plus amine" routes as they lead to significant amounts of by-product. Benzyl bromide with ammonia and methyl iodide with benzyl amine would both be unwise and wasteful. The routes illustrated are those of choice.

Second Examination (E) One Hour

For each of the questions 1-10 choose the set of reagents which <u>best</u> accomplishes the desired conversion.

1. 2-butanone ———⟶ 2-butylamine
   (a) sodium amide
   (b) 1. ammonia, 2. hydrogen, palladium on charcoal
   (c) 1. sodium cyanide, 2. sulfuric acid
   (d) 1. sodium bisulfate, 2. ammonia
   (e) sulfuric acid, sodium cyanide

2. butyric acid ———⟶ ethyl butyrate
   (a) ethanol, potassium hydroxide
   (b) 1. thionyl chloride, 2. ethanol
   (c) sulfuric acid, heat
   (d) 1. ammonia, 2. ethanol
   (e) 1. potassium hydroxide, 2. ethyl chloride

3. acetone ———⟶ monobromoacetone
   (a) hydrogen bromide, sulfuric acid
   (b) bromine, potassium hydroxide
   (c) phosphorus tribromide, bromine
   (d) bromine, carbon tetrachloride
   (e) bromine, acetic acid

4. cyclohexylamine ———⟶ cyclohexylmethylamine
   (a) methyl iodide, heat
   (b) 1. sulfuric acid, 2. methyl iodide
   (c) 1. phosphorus trichloride, 2. methyl lithium
   (d) 1. formaldehyde, 2. formic acid
   (e) 1. methanol, thionyl chloride, 2. water

5. benzene ———⟶ p-nitrobromobenzene
   (a) 1. bromine, iron, 2. sulfuric acid, nitric acid
   (b) 1. bromine, iron, 2. sodium nitrite, sulfuric acid
   (c) 1. sulfuric acid, nitric acid, 2. bromine, iron
   (d) bromine, nitric acid
   (e) 1. bromine, 2. nitric acid, acetic acid

6. benzene ———⟶ iodobenzene
   (a) iodine, iron
   (b) 1. chlorine, iron, 2. sodium iodide, acetone
   (c) 1. sulfuric acid, nitric acid, 2. tin, hydrogen chloride, sodium nitrite, sulfuric acid, 3. potassium iodide
   (d) iodine, sulfuric acid
   (e) phosphorus triiodide

7. cyclohexylamine ———> cyclohexene
   (a) potassium hydroxide, heat
   (b) 1. excess methyl iodide, 2. silver oxide, heat
   (c) 1. sulfuric acid, sodium nitrite, heat, 2. water
   (d) sulfuric acid, heat
   (e) 1. hydrazine, 2. potassium hydroxide, heat

8. ethyl bromide ———> propiophenone
   (a) 1. sodium cyanide, DMSO, 2. phenyl Grignard, 3. water
   (b) sodium hydroxide, acetophenone
   (c) 1. magnesium, ether, 2. acetophenone
   (d) 1. formaldehyde, sodium hydroxide, 2. bromobenzene, water
   (e) 1. formaldehyde, sodium hydroxide, 2. aluminum chloride, benzene

9. acrolein ———> 2,4-pentadienoic acid
   (a) cyclopentadiene, potassium permanganate
   (b) 1. ethyl Grignard, ether, 2. carbon dioxide
   (c) 1. sodium hydroxide, 2. sulfuric acid, heat
   (d) sodium acetate, acetic anhydride
   (e) sodium cyanide, water, 2. ethyl Grignard, ether

10. 1-bromobutane ———> n-butylamine
    (a) sodium amide
    (b) 1. sodium cyanide, 2. lithium aluminum hydride
    (c) 1. sulfuric acid, sodium nitrite, 2. tin, hydrogen chloride
    (d) 1. potassium phthalimide, 2. sodium hydroxide, water
    (e) ammonia

For each of questions 11-20 choose the major organic product(s) in the reaction indicated.

11. diethyl adipate + sodium ethoxide ———>
    (a) 2-carboethoxycyclopentanone
    (b) no reaction
    (c) cyclohexanone
    (d) ethyl hexanoate
    (e) sodium adipate

12. ethyl bromoacetate + zinc + 2-butanone ———>
    (a) 2-butanol
    (b) ethyl 3-hexenoate
    (c) ethyl 3-methyl-3-hydroxypentanoate
    (d) 5-methyl-4-hepten-3-one
    (e) ethyl acetate

13. 1,3-butadiene + heat ———>
    (a) no reaction
    (b) 1,3,5,7-octatetraene
    (c) 4-vinylcyclohexene
    (d) cyclooctatetraene

(e) ethylene

14. m-dinitrobenzene + ammonium polysulfide ⟶
   (a) m-diaminobenzene
   (b) m-nitrothiophenol
   (c) m-nitroaniline
   (d) di(m-nitrophenyl)amine
   (e) no reaction

15. α-naphthylamine + 1. sulfuric acid, sodium nitrite, 2. cuprous bromide ⟶
   (a) naphthalene
   (b) α-naphthyl bromide
   (c) α-naphthyl sulfate
   (d) no reaction
   (e) β-naphthyl bromide

16. sodio diethyl malonate + benzaldehyde ⟶
   (a) diethyl α-phenylmalonate
   (b) ethyl 3-phenylpropionate
   (c) ethyl 2-carboethoxy-3-phenylpropenoate
   (d) no reaction
   (e) sodium benzylate

17. propionitrile + lithium aluminum hydride ⟶
   (a) n-butylamine
   (b) n-butyramide
   (c) butane
   (d) butyric acid
   (e) n-propylamine

18. β-isopropylnaphthalene + 1. oxygen, 2. aqueous acid ⟶
   (a) naphthalene
   (b) β-isopropenylnaphthalene
   (c) β-naphthoic acid
   (d) β-naphthol
   (e) α-naphthol

19. N,N-dimethylaniline + sulfuric acid + sodium nitrite ⟶
   (a) benzenediazonium sulfate
   (b) p-nitroso-N,N-dimethylaniline
   (c) N-methyl-N-nitrosoaniline
   (d) aniline
   (e) benzene

20. methylenetriphenylphosphine + 2-butanone ⟶
   (a) tri-2-butylphosphine
   (b) 2-methyl-1-butene
   (c) formaldehyde
   (d) no reaction
   (e) 2-butanol

For each of questions 21-25 choose the reagent which will serve to differentiate most quickly and clearly the indicated compounds.

21. aniline and N-methylaniline
    (a) 1. p-toluenesulfonyl chloride, 2. aqueous sodium hydroxide
    (b) acetyl chloride
    (c) 1. benzaldehyde, 2. bromine
    (d) bromine, carbon tetrachloride
    (e) acetic acid

22. p-methylphenol and benzyl alcohol
    (a) iodine
    (b) sodium
    (c) zinc chloride, hydrogen chloride
    (d) potassium permanganate
    (e) dilute aqueous sodium hydroxide

23. p-methylphenol and benzoic acid
    (a) aqueous sodium bicarbonate
    (b) dilute aqueous sodium hydroxide
    (c) iodine
    (d) sodium
    (e) aluminum chloride

24. 3-pentanone and pentanal
    (a) iodine, potassium hydroxide
    (b) sodium bisulfite
    (c) hydrazine
    (d) sodium cyanide
    (e) aniline

25. 2,4,6-trinitrophenol and phenol
    (a) iodine
    (b) sodium
    (c) aqueous sodium bicarbonate
    (d) potassium permanganate
    (e) dilute aqueous sodium hydroxide

First Examination (E)   Answer Set

1.(b) 2.(b) 3.(e) 4.(d) 5.(a) 6.(c) 7.(b) 8.(a) 9.(d) 10.(d) 11.(a) 12.(c)
13.(c) 14.(c) 15.(d) 16.(c) 17.(a) 18.(d) 19.(b) 20.(b) 21.(a) 22.(e)
23.(a) 24.(b) 25.(c)

Third Examination (A)   One Hour

1. (25 pts, 15 min) Show the fundamentals of the modified Skraup synthesis of quinoline using the reagents given below. The involvement of each reagent should be demonstrated.

   HOCH$_2$-CH(OH)-CH$_2$OH   H$_3$PO$_4$ C$_6$H$_5$-NH$_2$

2. (25 pts, 15 min) Predict the major organic product in each of the following reactions. Illustrate the proper stereochemistry where appropriate.

   (a) m-O$_2$N-C$_6$H$_4$-CH(CH$_3$)$_2$ $\xrightarrow{O_2, H_3O^+}$

   (b) methyl α-D-glucopyranoside $\xrightarrow[\text{2. aq. acid}]{\text{1. excess } IO_4^-}$

   (c)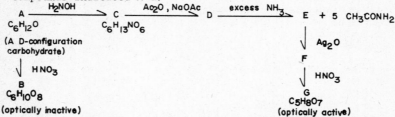

   (followed by treatment with ethanolic hydrogen chloride)

   (d) C$_6$H$_5$-CHO + large excess diethyl malonate $\xrightarrow{\text{EtOH, NaOEt}}$

3. (25 pts, 15 min) Give correct structures (including stereochemistry) for compounds indicated A–G.

   A $\xrightarrow{H_2NOH}$ C $\xrightarrow{Ac_2O, NaOAc}$ D $\xrightarrow{\text{excess } NH_3}$ E + 5 CH$_3$CONH$_2$

   C$_6$H$_{12}$O   C$_6$H$_{13}$NO$_6$

   (A D-configuration carbohydrate)

   ↓ HNO$_3$ ↓ Ag$_2$O
   B   F
   C$_6$H$_{10}$O$_8$ ↓ HNO$_3$
   (optically inactive)   G
                                                       C$_5$H$_8$O$_7$
                                                       (optically active)

4. (25 pts, 15 min) Devise syntheses for each of the following compounds using the indicated starting compound and any other organic and inorganic reagents thought necessary.
   (a) D-erythrose from D-glyceraldehyde
   (b) alanine from acetaldehyde
   (c) quinoline-4-carboxylic acid from aniline
   (d) HOOC-CH$_2$-CH$_2$-CH$_2$-COOH from EtOOC-CH$_2$-COOEt

Third Examination (A)   Answer Set

1. [Mechanism scheme showing the Skraup-type synthesis of quinoline from glycerol and aniline with H₃PO₄, proceeding through protonation/dehydration of glycerol to acrolein, Michael addition of aniline, cyclization, dehydration, and oxidation to give quinoline.]

The difference between this and the original Skraup synthesis is the use of phosphoric acid in place of the sulfuric acid and nitrobenzene. Phosphoric acid serves both as acid and oxidizing agent. In writing the mechanism one must be aware of the intermediacy of acrolein, generated from glycerol by a series of acid-base reactions, and the subsequent Michael-type addition of aniline to the acrolein. In the next stage the molecule undergoes an intramolecular electrophilic substitution reaction followed by acid-catalyzed dehydration and oxidation. An understanding of simple acid-base chemistry is invaluable to understanding complex reactions. One should also note that other bases present in the reaction mixture, such as aniline, may serve as proton acceptors rather than $H_2PO_4^-$. This has been used throughout the mechanism as written simply to point out the catalytic function of the acid reactant.

2. (a) [Structure: 3-nitrophenol, HO-C₆H₄-NO₂]

This is another example of the cumene synthesis of phenols.

(b)
$$\begin{array}{c} CHO \\ H-C-OH \\ CH\ OH \end{array} + CH_3OH + HCOOH + \begin{array}{c} CHO \\ | \\ CHO \end{array} \xrightarrow[\text{2. aq. acid}]{\text{1. excess } IO_4^-} \text{[methyl glycoside structure with OCH}_3\text{]}$$

For this standard degradation of a glycoside it is important that one recognize from the name of the compound that it is a <u>methyl</u> glycoside. It is derived from an <u>aldohexose</u> of D-configuration, and it contains a <u>six-membered</u> ring. Other stereochemical subtleties are of no consequence. In the initial reaction all vicinal diol linkages are cleaved (there are only two) yielding formic acid and the complex acetal

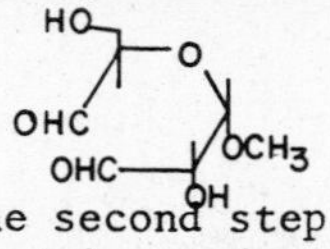

which is hydrolyzed in the second step. While carbohydrate nomenclature may seem (to beginning organic students and many of their instructors as well) to be as obscure as IRS regulations, it is quite systematic and requires the memorization of only a few simple rules.

(c)

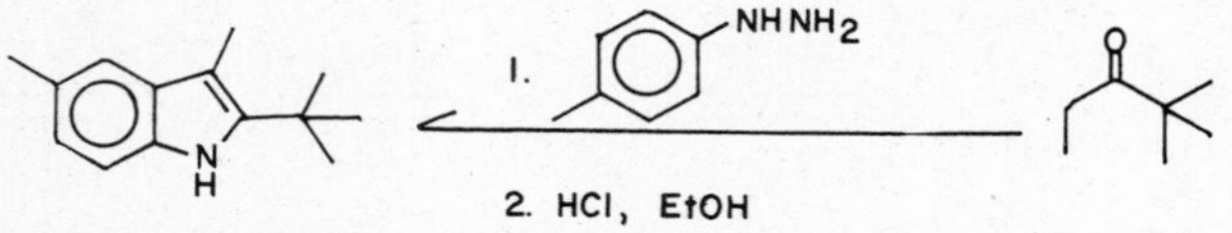

The initial reaction between the carbonyl compound and the substituted hydrazine should be recognized readily as leading to a carbon-nitrogen doubly bonded system, the phenylhydrazone. This compound is then subject, under acidic conditions, to the Fischer indole synthesis. In the latter reaction the original carbonyl becomes the 2-position of the indole (numbering shown below) and the α-carbon, bearing active hydrogens, becomes the 3-position. Substituents on these carbons are consequently located at the 2- and 3-positions, respectively, of the indole ring system. The methyl para- to the hydrazine linkage of course retains this relationship in the product.

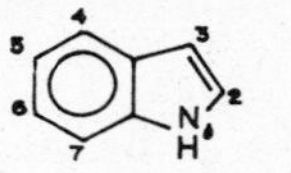

(d)

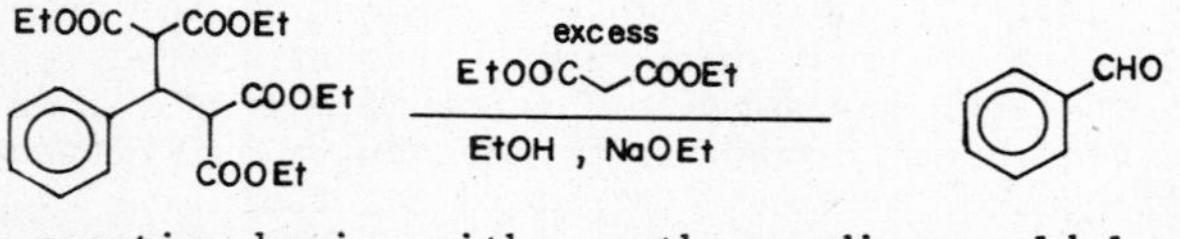

This reaction begins with a rather ordinary aldol-related condensation with accompanying dehydration of adduct to yield

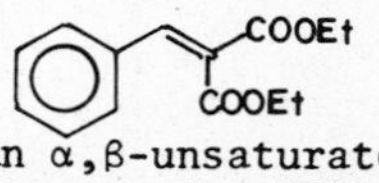

This intermediate, an α,β-unsaturated ester, then undergoes Michael addition of a second anion from diethyl malonate generating the product shown.

3.

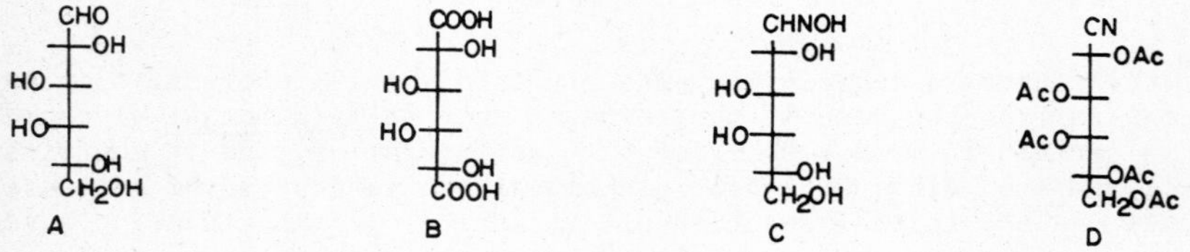

A: CHO, —OH, HO—, HO—, —OH, CH$_2$OH

B: COOH, —OH, HO—, HO—, —OH, COOH

C: CHNOH, —OH, HO—, HO—, —OH, CH$_2$OH

D: CN, —OAc, AcO—, AcO—, —OAc, CH$_2$OAc

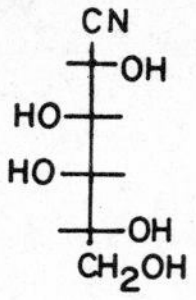

```
      CN                    CHO                   COOH
    ──OH                  ──                    ──
HO──                   HO──                  HO──
HO──                   HO──                  HO──
    ──OH                  ──OH                  ──OH
    CH₂OH                 CH₂OH                 COOH
      E                     F                     G
```

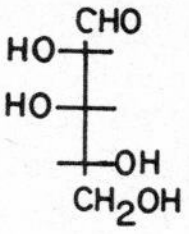

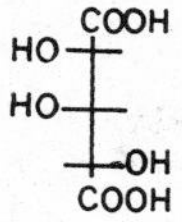

There are two fundamental aspects to this problem, understanding the stereochemistry, and knowing the reactions. The reactions are involved with shortening the carbohydrate chain by one carbon (Wohl degradation), that is, conversion of A to F, and the oxidation of the termini of the carbohydrates to carboxylic acid functions (G and B). Clues to the stereochemistry are found in these reactions. Thus, as compound B is optically inactive, it must be a meso- structure, of which two are possible as shown below, B and B'. Of these B' is found to be unreasonable in light of the data for compound G (we are told it is optically active).

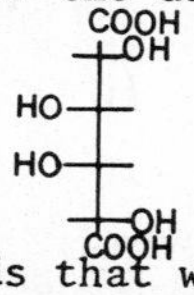

```
      COOH                    COOH
    ──OH                    ──OH
HO──                      ──OH
HO──                      ──OH
    ──OH                    ──OH
    COOH        B           COOH        B'
```

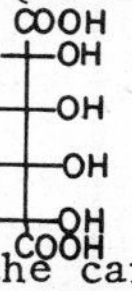

The diacids that would result from the carbohydrate derived from A by the Wohl degradation would be G and G'. Of these only G is optically active. Thus one deduces the stereochemistry of all intermediate structures.

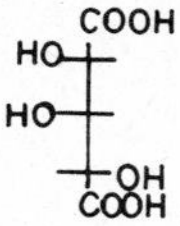

```
      COOH                    COOH
HO──                        ──OH
HO──                        ──OH
    ──OH                    ──OH
    COOH                    COOH
      G                       G'
```

4. (a)
```
  CHO          OC─O              COOH            CN           CN                  
H─COH   Na(Hg) H─COH    Ba(OH)₂  H─COH    H₃O⁺   H─COH   +   HOC─H    NaCN    CHO
H─COH    ──►   H─COH     ◄────   H─COH    ──►   H─COH        H─COH   ─────  H─COH
CH₂OH    H⁺    H₂C─┘             CH₂OH           CH₂OH        CH₂OH   NaHSO₄ CH₂OH
```

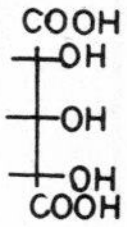

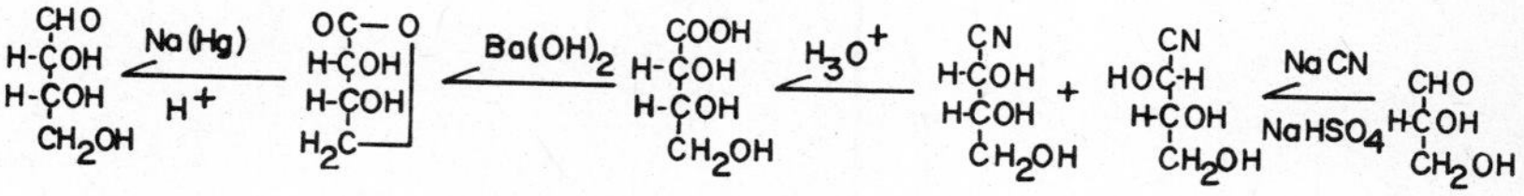

The Killiani-Fischer synthetic method is used here for the extension of a carbohydrate chain by a single carbon. In the first reaction, diastereoisomers are formed, and must be separated; one of the diastereoisomers is used for the subsequent reactions. In practice it might be easier to hydrolyze the reaction mixture first and then separate the acids.
But how does one know which of the diastereoisomers is the one desired? This could be determined by oxidation of a sample of each with nitric acid; the resulting diacid from the compound leading to D-erythrose would be optically inactive.

(b)

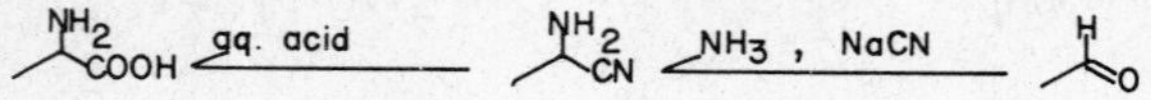

$\underset{\text{COOH}}{\overset{NH_2}{\diagup}} \xleftarrow{\text{aq. acid}} \underset{\text{CN}}{\overset{NH_2}{\diagup}} \xleftarrow{NH_3, \; NaCN} \underset{O}{\overset{H}{\diagup}}$

This is an example of the Strecker synthesis of amino acids. The "R" group (in biological terms) of the amino acid is included in the starting aldehyde. From this is formed the nitrogen analogue of a cyanohydrin, which is hydrolyzed to the target molecule.

(c) 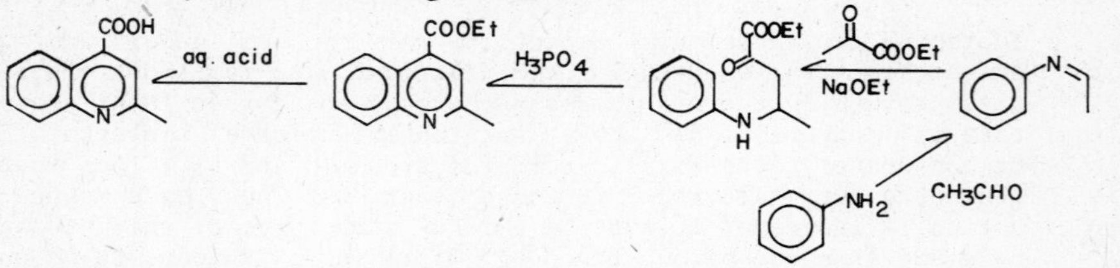 [quinoline-COOH] $\xleftarrow{\text{aq. acid}}$ [quinoline-COOEt] $\xleftarrow{H_3PO_4}$ [aniline-CH(CH_3)-CH_2-COOEt with ketone COOEt] $\xleftarrow{NaOEt}$ [PhN=CH–] + [PhNH_2] + CH_3CHO

The route shown here for quinoline ring synthesis is one of several of importance for the laboratory synthesis of natural alkaloids. Ring closure is by intramolecular electrophilic aromatic substitution.

(d) 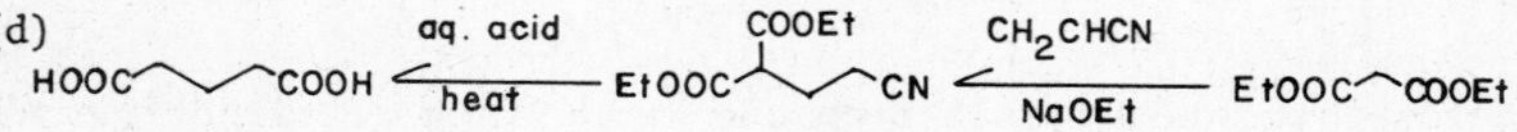 $HOOC\!\frown\!\!\frown\!COOH \xleftarrow[\text{heat}]{\text{aq. acid}} EtOOC\!\frown\!\!\frown\!CN \xleftarrow[NaOEt]{CH_2CHCN} EtOOC\!\frown\!COOEt$

The route shown uses a Michael addition to acrylonitrile, followed by hydrolysis and decarboxylation. Enterprising students may devise (and have devised) other routes, usually much longer.

Third Examination (B)   One Hour

1. (25 pts, 15 min) Predict the products of quantitative periodate oxidation (after acid hydrolysis) for each of the following compounds:

   (a) methyl α-D-ribofuranoside
   (b) methyl β-D-glucopyranoside
   (c) methyl 2,3,5,6-tetra-O-methyl-α-D-glucopyranoside

2. (25 pts, 15 min) Give the correct structures for compounds A-G.

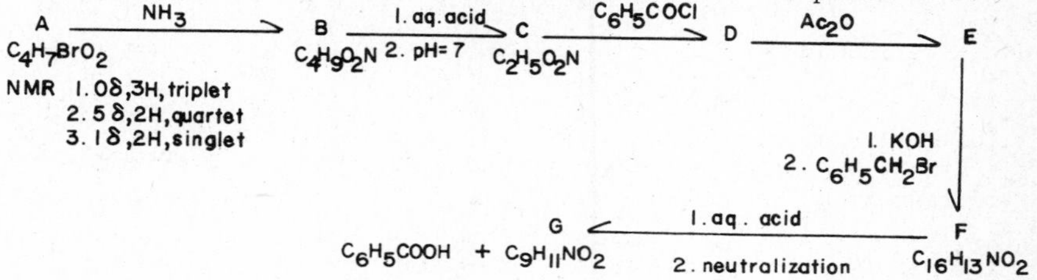

   A $\xrightarrow{NH_3}$ B $\xrightarrow[2.\ pH=7]{1.\ aq.\ acid}$ C $\xrightarrow{C_6H_5COCl}$ D $\xrightarrow{Ac_2O}$ E

   $C_4H_7BrO_2$ $C_4H_9O_2N$ $C_7H_5O_2N$

   NMR 1.0 δ, 3H, triplet
       2.5 δ, 2H, quartet
       3.1 δ, 2H, singlet

   E $\downarrow$ 1. KOH
     2. $C_6H_5CH_2Br$

   G $\xleftarrow[2.\ neutralization]{1.\ aq.\ acid}$ F

   $C_6H_5COOH + C_9H_{11}NO_2$ $C_{16}H_{13}NO_2$

3. (25 pts, 15 min) Devise syntheses for any two of the following compounds, starting with the indicated materials and using any other organic and inorganic reagents thought necessary.

   (a) 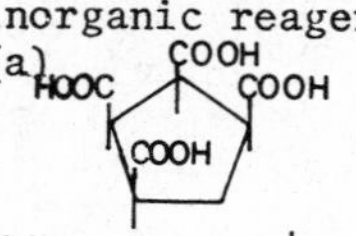 [furanose triacid with HOOC, COOH, COOH] from [maleic anhydride]

   (b) 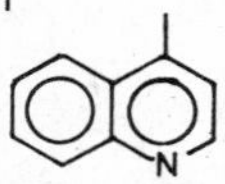 [4-methylquinoline] from [acrolein / methyl vinyl ketone]

   (c) 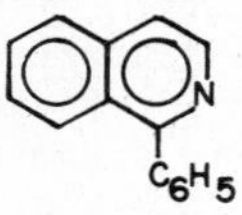 [1-phenylisoquinoline with C_6H_5] from benzoyl chloride

   (d)  HOOC~~COOH from [acrolein]

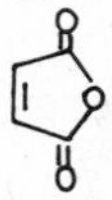

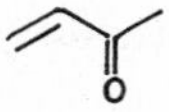

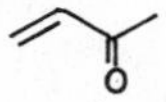

4. (25 pts, 15 min) Show all steps in the synthesis of D-ribose and D-glyceraldehyde from D-erythrose. Indicate at which points more than one product would be obtained and which of these would be required to continue the synthesis. The following structures are given:

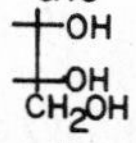

   D-erythrose: CHO, OH, OH, CH₂OH
   D-ribose: CHO, OH, OH, OH, CH₂OH
   D-glyceraldehyde: CHO, OH, CH₂OH

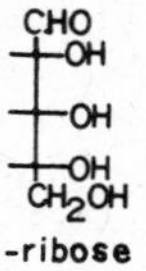

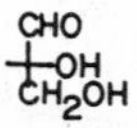

Third Examination (B)   Answer Set

1. (a) $CH_3OH + HC(=O)-CH(=O) + H-C(=O)-OH$ with CH_2OH ⟵

   (b) $CH_3OH + HC(=O)-CH(=O) + HCOOH + H-C(=O)-OH$ with CH_2OH ⟵

   (c) no oxidation ⟵

The major points of recognition in the names relate to size of the carbohydrate, D- or L-configuration, and ring size. Other points are unimportant in answering this question.

2. $BrCH_2COC_2H_5$ $H_2NCH_2COC_2H_5$ H_2NCH_2COOH PhC(=O)NH-CH$_2$-COOH
      A   B   C   D

   E   F   G

In this problem the structural information is not so obvious as previously. Given the nmr data for compound A, the beginning student might come up with either of two structures, A or A'.

$BrCH_2COC_2H_5$ $CH_3CH_2COCH_2Br$
A   A'

While A' might seem strange on a chemical basis, it is rigorously excluded (to the beginning student) only with the generation of compound C. Even if a reaction such as that needed to form B could proceed from A' without fragmentation, hydrolysis could not lead to a two-carbon product. The remainder of the synthesis is the azlactone synthesis of amino acids proceeding through hippuric acid, compound E. The similarity between E and a malonic ester should be noted. Like a malonic ester, E can be alkylated, and in this case ultimate hydrolysis leads to an amino acid.

3. (a) HOOC-[cyclopentane with COOH, COOH, COOH] ⟵ 1. KMnO$_4$ / 2. aq. acid ⟵ [norbornene anhydride] ⟵ [cyclopentadiene] + [maleic anhydride]

The target molecule here obviously must be approached using reactions of high stereoregularity. Two potential carboxylic acid linkages are already in a *cis* relationship (in the maleic anhydride) if a *cis* addition to the olefinic linkage is accomplished. A suitable reaction is the Diels-Alder reaction. Using the proper diene (cyclopentadiene) one provides a five-membered ring system and a function (the olefinic linkage) that can be converted into the remaining two carboxylic acids with the proper stereochemistry. The solution to this problem requires experience in looking at a target molecule and enumerating the ways in which the functional groups present may be generated.

(b) quinoline ⟵[H_3PO_4] aniline + acrolein

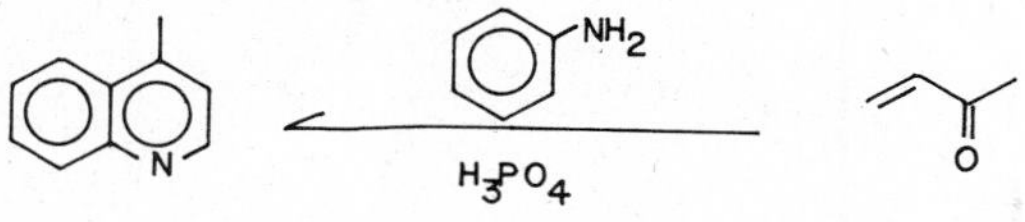

This is a modification of the Skraup synthesis in which the unsaturated carbonyl compound is added directly rather than generated *in situ*.

(c) 1-phenylisoquinoline ⟵[Pd, N_2, heat] 3,4-dihydro-1-phenylisoquinoline ⟵[P_2O_5] N-benzoyl-β-phenethylamine ⟵ β-phenethylamine + benzoyl chloride (COCl)

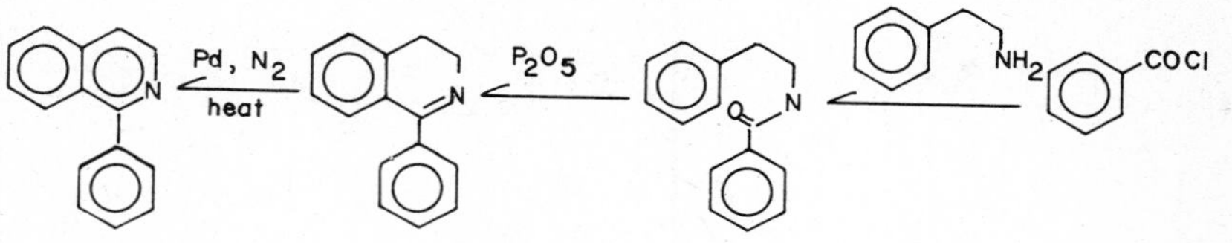

This synthesis is an adaptation of the classic Bischler-Napieralski method for the preparation of an isoquinoline ring system. The acid-catalyzed ring closure is performed here with phosphorus pentoxide and aromatization by dehydrogenation over a standard hydrogenation catalyst, the product hydrogen being swept away with nitrogen.

(d) HOOC~~~COOH ⟵[Br_2, KOH / H_2O] HOOC~~~C(=O)CH₃ ⟵[H_3O^+, heat] EtOOC~~~C(=O)CH₂CH₂COOEt ⟵[EtOOC-CH(COOEt), NaOEt] CH₂=CH-C(=O)CH₃

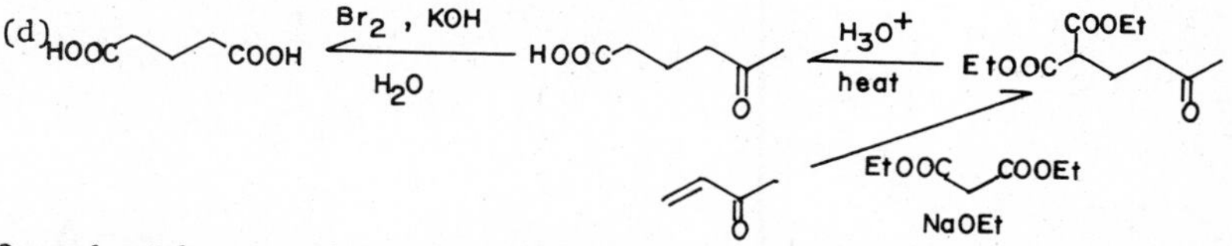

One should recognize that generation of a carboxylic acid terminus is required from a methyl ketone. This is accomplished by the haloform reaction. One should also recognize that malonic ester may be used in Michael-type additions.

4. Illustrated here are the standard routes for the accomplishment of the desired conversions, the Kiliani synthesis for D-ribose and the Wohl degradation for D-glyceraldehyde. The Ruff degradation would be equally acceptable. Separations in the Kiliani synthesis can be performed either at the cyanohydrin stage or after the next step, hydrolysis to the acids.

D-ribose synthesis:

$$\begin{array}{c}\text{CHO}\\\text{|--OH}\\\text{|--OH}\\\text{CH}_2\text{OH}\end{array} \xrightarrow{\text{NaCN},\ H_2O} \begin{array}{c}\text{CN}\\\text{|--OH}\\\text{|--OH}\\\text{|--OH}\\\text{CH}_2\text{OH}\end{array} + \begin{array}{c}\text{CN}\\\text{HO--|}\\\text{|--OH}\\\text{|--OH}\\\text{CH}_2\text{OH}\end{array} \xrightarrow{\text{aq. acid}} \begin{array}{c}\text{COOH}\\\text{|--OH}\\\text{|--OH}\\\text{|--OH}\\\text{CH}_2\text{OH}\end{array} \xrightarrow{\text{Ba(OH)}_2} \begin{array}{c}\text{OC--O}\\\text{|--OH|}\\\text{|--OH}\\\text{|--OH}\\\text{CH}_2\text{OH}\end{array} \xrightarrow[\text{aq. acid}]{\text{Na(Hg)}} \begin{array}{c}\text{CHO}\\\text{|--OH}\\\text{|--OH}\\\text{|--OH}\\\text{CH}_2\text{OH}\end{array}$$

SEPARATE

D-glyceraldehyde synthesis:

$$\begin{array}{c}\text{CHO}\\\text{|--OH}\\\text{|--OH}\\\text{CH}_2\text{OH}\end{array} \xrightarrow{\text{H}_2\text{NOH}} \begin{array}{c}\text{CHNOH}\\\text{|--OH}\\\text{|--OH}\\\text{CH}_2\text{OH}\end{array} \xrightarrow{\text{Ac}_2\text{O}} \begin{array}{c}\text{CN}\\\text{|--OAc}\\\text{|--OAc}\\\text{CH}_2\text{OAc}\end{array} \xrightarrow{\text{NH}_3} \begin{array}{c}\text{CN}\\\text{|--OH}\\\text{|--OH}\\\text{CH}_2\text{OH}\end{array} \xrightarrow{\text{Ag}_2\text{O}} \begin{array}{c}\text{CHO}\\\text{|--OH}\\\text{CH}_2\text{OH}\end{array}$$

Third Examination (C)   One Hour

1. (25 pts, 15 min) Write reactions involving real compounds (that is, actual structures, <u>not</u> R-, etc.) of six carbon atoms or less that illustrate the use of each <u>of</u> the following reagents. Products must be isolable, and any pertinent stereochemistry must be indicated.
   (a) CH_3COOH, H_2O_2
   (b) $Ba(ClO_3)_2$, OsO_4
   (c) KIO_4
   (d) PBr_3, Br_2
   (e) Zn

2. (20 pts, 10 min) The tripeptide alanylglycylproline was suspected of being a fragment in the degradative analysis of a much larger polypeptide. To prove this, an independent synthesis was performed. Give details of a synthesis of this tripeptide starting with the free amino acids and any other reagents thought necessary.

3. (25 pts, 15 min) Show all reagents (and isolable intermediate products) involved in the following conversions.
   (a)
   (b)
   (c)

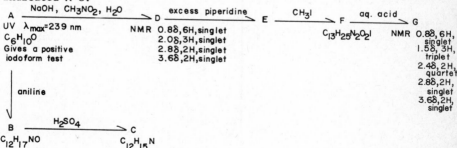

4. (30 pts, 20 min) Give the correct structure for each of the compounds indicated A-G.

   A $\xrightarrow{NaOH, CH_3NO_2, H_2O}$ D $\xrightarrow{excess\ piperidine}$ E $\xrightarrow{CH_3I}$ F $\xrightarrow{aq.\ acid}$ G

   A: UV λ_{max}=239 nm, $C_6H_{10}O$, Gives a positive iodoform test

   D: NMR 0.8δ, 6H, singlet
          2.0δ, 3H, singlet
          2.8δ, 2H, singlet
          3.6δ, 2H, singlet

   F: $C_{13}H_{25}N_2O_2I$

   G: NMR 0.8δ, 6H, singlet
          1.5δ, 3H, triplet
          2.4δ, 2H, quartet
          2.8δ, 2H, singlet
          3.6δ, 2H, singlet

   A $\xrightarrow{aniline}$ B $\xrightarrow{H_2SO_4}$ C

   B: $C_{12}H_{17}NO$
   C: $C_{12}H_{15}N$

Third Examination (C)   Answer Set

1. (a) CH$_2$=CHCH$_3$ $\xrightarrow{\text{CH}_3\text{COOH}, \text{H}_2\text{O}_2}$ (epoxide from propene)

   (b) CH$_2$=CHCH$_3$ $\xrightarrow[\text{OsO}_4]{\text{Ba(ClO}_3)_2}$ H$_3$C–CH(OH)–CH(OH)–CH$_3$ (meso diol drawn with wedges, CH$_3$ groups)

   (c) (CH$_3$)$_2$C(OH)–C(OH)(CH$_3$)$_2$ $\xrightarrow{\text{KIO}_4, \text{H}_2\text{O}}$ 2 (CH$_3$)$_2$CO

   (d) (CH$_3$)$_2$CHCOOH $\xrightarrow{\text{PBr}_3, \text{Br}_2}$ (CH$_3$)$_2$CBrCOOH

   (e) BrCH$_2$COOCH$_3$ $\xrightarrow[\text{2. H}_2\text{O}]{\text{1. Zn, acetone}}$ (CH$_3$)$_2$C(OH)CH$_2$COOCH$_3$

2. CH$_3$CH(NH$_2$)COOH $\xrightarrow{\text{C}_6\text{H}_5\text{CH}_2\text{OCOCl}}$ C$_6$H$_5$CH$_2$O–CO–NH–CH(CH$_3$)–COOH $\xrightarrow{\text{PCl}_5}$ C$_6$H$_5$CH$_2$O–CO–NH–CH(CH$_3$)–COCl

   $\xrightarrow{\text{glycine}}$ C$_6$H$_5$CH$_2$O–CO–NH–CH(CH$_3$)–CO–NH–CH$_2$–COOH $\xrightarrow{\text{PCl}_5}$ C$_6$H$_5$CH$_2$O–CO–NH–CH(CH$_3$)–CO–NH–CH$_2$–COCl

   $\xrightarrow{\text{proline}}$ C$_6$H$_5$CH$_2$O–CO–NH–CH(CH$_3$)–CO–NH–CH$_2$–CO–(proline) $\xrightarrow{\text{H}_2, \text{Pd}}$ H$_2$N–CH(CH$_3$)–CO–NH–CH$_2$–CO–(proline-COOH)

   Other methods of activation for coupling may be used, for example, mixed anhydride, active ester, or azide. Any of these would be suitable here. The amino terminus may also be protected with other functions, for example, the t-BOC group.

3. (a) HOOC(CH$_2$)$_4$COOH $\xleftarrow{\text{KMnO}_4}$ cyclohexene $\xleftarrow{\text{H}_2\text{SO}_4}$ cyclohexanol $\xleftarrow[\text{LiAlH}_4]{\text{NaBH}_4 \text{ or}}$ cyclohexanone

   One should realize that the same number of carbon atoms are present in the starting material as in the target molecule. Thus the fundamental reaction type required is carbon-carbon bond cleavage. An olefinic linkage is convenient for this and must be generated from the ketone in several steps.

(b)

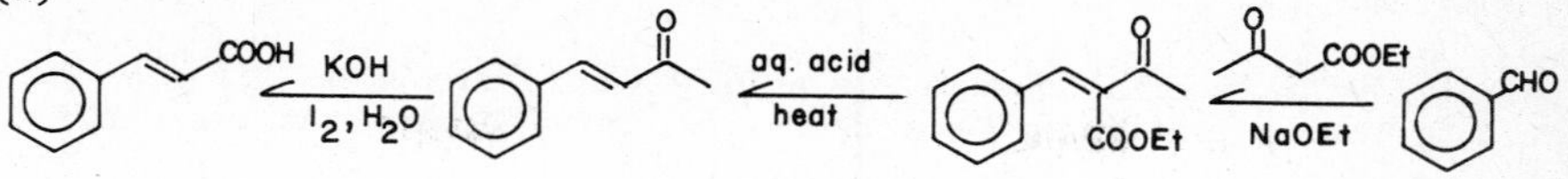

[Reaction scheme: cinnamic acid derivative → via KOH, I₂, H₂O → styryl ketone → aq. acid, heat → phenyl diketone with COOEt groups → NaOEt → benzaldehyde with CHO and COOEt]

The solution requires the combination of several types of reactions. It is important that one be able to look at the target molecule and see that portion which is derived from the starting material, that is, the carbonyl carbon and its adjacent carbon.

(c) In this conversion shown below two points are of importance. First, a reaction is needed that generates a six-membered ring. Given the site of unsaturation within it, the Diels-Alder reaction becomes the obvious choice. Second, given the stereochemistry of the reaction, a cis- to trans- interconversion of the alkene is required. Two routes are shown. The one proceeding through the acetylene dicarboxylic acid is probably most familiar, although the other is also convenient and reasonable.

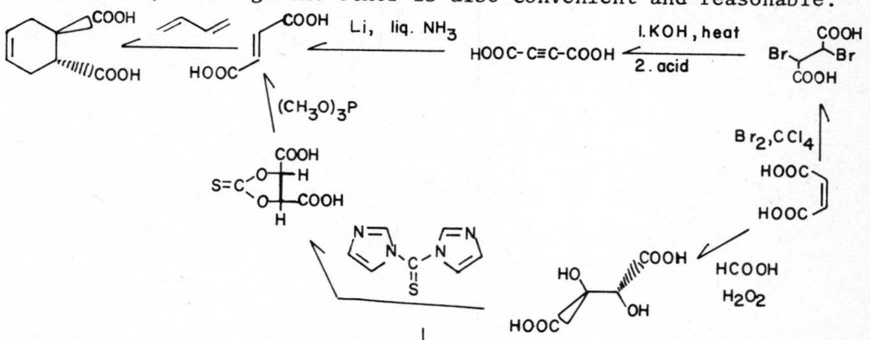

[Reaction scheme showing cyclohexene dicarboxylic acid synthesis via Diels-Alder with butadiene and HOOC-C≡C-COOH, Li/liq. NH₃, (CH₃O)₃P, KOH/heat, Br₂/CCl₄, HCOOH/H₂O₂ routes]

4. 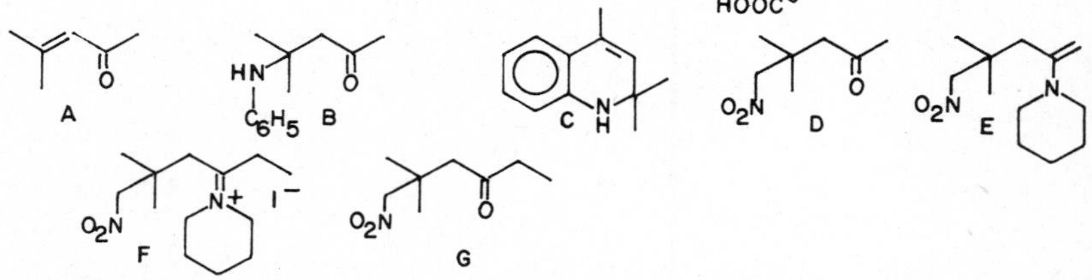 [Structures labeled A through G: A (methyl vinyl ketone derivative), B (amine with C₆H₅), C (quinoline derivative), D (O₂N ketone), E (O₂N piperidine enamine), F (O₂N iminium salt with I⁻), G (O₂N ethyl ketone)]

There are several important points to recognize here as the fundamental structural data is scattered. The UV data for A should give the fundamental β-disubstituted α,β-unsaturated ketone structure. The immediate reactions using this species should be clear. Difficulty arises in a decision on the structures of E and F, as two possibilities are available for each. The proper one in each case is found by an analysis of the nmr data for G which indicates an ethyl ketone.

Third Examination (D)   One Hour

1. (25 pts, 15 min) Predict the major organic product(s) in each of the following reactions. Be sure to indicate the proper stereochemistry.

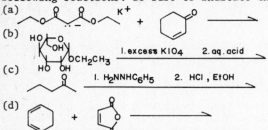

2. (25 pts, 15 min) Show in detail the reasoning, all chemical reactions, and all physical measurements in the determination of the absolute configuration of D-threose. Assume that the absolute configuration of D-glyceraldehyde is known.

3. (25 pts, 15 min) Give details for the synthesis of each of the following compounds from the indicated starting materials using any other reagents thought necessary.

   (a) [quinoline] from [p-aminoaniline/p-toluidine structure]

   (b) cinnamic acid from diethyl malonate

   (c) methyl butyl ether from methyl vinyl ketone

4. (25 pts, 15 min) Give the correct structure for each of the lettered compounds A–H.

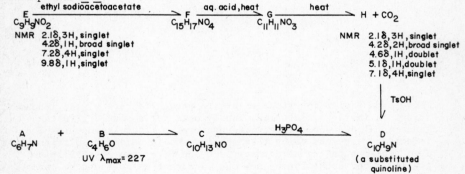

Third Examination (D)   Answer Set

1. (a) Cyclohexene with CH(COOMe)$_2$ substituent, and epoxide (O) on ring.

   (b) CH$_3$CH$_2$OH + HC(=O)-C(=O)H + HCOOH + HOCH$_2$-CH(OH)-CHO (structure: CHO, HO-C-H, CH$_2$OH)

   (c) 2-methylindole + 2-ethylindole

   (d) norbornene-fused anhydride (endo Diels–Alder adduct of cyclopentadiene with maleic anhydride)

2. Two set of reactions are required, knowing that one is dealing with a tetrose, as may be determined by either of several means of molecular weight measurement. The first of these is an oxidation with nitric acid to generate the dicarboxylic acid.

   CHO   COOH
   HO—H   HNO$_3$ HO—H
   H—OH   ———→   H—OH
   CH$_2$OH   COOH

   The critical observation to be made is that the product diacid is optically active. Thereby the structure <u>must</u> be either that shown or its enantiomer. To determine which a degradation must be performed. Comparison of the glyceraldehyde product with an authentic sample establishes the absolute structure of the starting carbohydrate.

   CHO   H$_2$NOH   CHNOH   Ac$_2$O   CN   NH$_3$ CN   Ag$_2$O
   HO—H   ———→   HO—H   ———→   AcO—H   ———→   HO—H   ———→   CHO
   H—OH   H—OH   H—OAc   H—OH   H—OH
   CH$_2$OH   CH$_2$OH   CH$_2$OAc   CH$_2$OH   CH$_2$OH

3. (a) p-toluidine (ArNH$_2$) + CH$_3$COCH$_2$COCH$_3$ ⟶ imine intermediate —H$_3$PO$_4$→ 2,4-dimethylquinoline (Doebner–Miller / Combes synthesis)

   (b) EtOOC-CH$_2$-COOEt + PhCHO —NaH→ PhCH=C(COOEt)$_2$ —aq. acid, heat→ PhCH=CHCOOH

In this type of problem one looks at the target molecule for that portion which is derived from the required starting material. For a malonic ester synthesis this is the carboxyl and adjacent carbons.

(c) CH₂=CH-CO- →(HCl, CH₃OH)→ H₃CO-CH₂-CH₂-CH(OCH₃)₂ →(aq. acid)→ O=CH-CH₂-CH₂-OCH₃ →(Sn, HCl)→ CH₃CH₂CH₂CH₂-OCH₃

The source of the butyl group for the target molecule is not immediately obvious. One should have some clue that it involves a Michael-type of addition reaction initially. The ketone may also be removed by a Wolff-Kischner reduction.

4.

A: aniline (C₆H₅NH₂)
B: CH₂=CH-CHO (acrolein)
C: 1,2,3,4-tetrahydroquinoline intermediate with CHO
D: methylquinoline
E: N-acetyl-2-acetyl aniline (o-aminoacetophenone acetamide)
F: 2-(acetyl-vinyl)-N-acetyl aniline with COOEt
G: 2-(acetyl-vinyl)aniline with COOH
H: 2-methyl-quinoline derivative

This problem is one of the more difficult presented here. The necessary data for structure elucidation must be gained by looking at several places. While the structure for A is obvious, the student may entertain more than one possibility for B. The point of methyl substitution on D depends on which possibility is correct. The nmr data for E might at first be confusing. One should keep in mind the presence of a disubstituted phenyl ring, definitely not para- and probably not meta-. The clinching point in determining structure E comes with the hydrolysis of F and the generation of acetic acid. With this data one can readily write the remaining structures.

Third Examination (E)   One Hour

For each of questions 1-7 choose the <u>major</u> organic product(s) in the reaction indicated.

1. p-methoxyaniline + acrolein + phosphoric acid ⟶
   (a) 2-methoxyquinoline
   (b) 1-methoxyisoquinoline
   (c) 5-methoxyquinoline
   (d) N-(p-methoxyphenyl)acrylamide
   (e) p-methoxyanilinium phosphate

2. methyl β-D-ribofuranoside + sodium hydroxide + water ⟶
   (a) methanol + D-ribose
   (b) methyl α-D-ribofuranoside
   (c) no reaction
   (d) methyl β-L-ribofuranoside
   (e) dimethyl ether + maltose

3. glycine + 1. benzoyl chloride, 2, acetic anhydride ⟶
   (a) N-benzoylglycine
   (b) hippuric acid
   (c) N-benzoylglycine acetate
   (d) phenylalanine
   (e) glycyl anhydride

4. D-ribose + excess phenylhydrazine ⟶
   (a) D-ribose phenylhydrazone
   (b) D-erythrose
   (c) D-erythrose osazone
   (d) β-D-ribose
   (e) D-arabinose osazone

5. cyclohexanone + 1. piperidine, 2. methyl iodide, 3. aqueous acid ⟶
   (a) N-methylpiperidine
   (b) methylcyclohexane
   (c) 2-methylcyclohexanone
   (d) 1-methylcyclohexanol
   (e) 1-methylcyclohexene

6. alanine + dicyclohexylcarbodiimide ⟶
   (a) alanylalanine + dicyclohexylurea
   (b) cyclohexyl alaninate
   (c) N-cyclohexylalanine
   (d) dicyclohexyl carbonate + alanylalanine
   (e) polyalanine + dicyclohexylurea

7. alanine + 1 equivalent benzyl chloroformate ⟶
   (a) N-carbobenzyloxyalanine
   (b) alanyl carbobenzyloxyl anhydride
   (c) benzyl alaninate
   (d) benzyl chloride
   (e) N-benzylalanine

For each question 8-14 choose the <u>best</u> set of reagents and/or conditions to effect the indicated conversion.

8. D-ribose ⟶ methyl α-D-ribofuranoside
   (a) p-toluenesulfonic acid, methanol
   (b) methyl iodide, sodium hydroxide
   (c) methanol, dicyclohexylcarbodiimide
   (d) diazomethane
   (e) methanol, sodium hydroxide

9. acetaldehyde ⟶ alanine
   (a) 1. ammonia, 2. carbon dioxide
   (b) 1. sodium cyanide, water, 2. sodium hydroxide
   (c) 1. sodium cyanide, ammonia, 2. aqueous acid
   (d) 1. sodium amide, 2. carbon dioxide
   (e) 1. sodium permanganate, 2. ammonia

10. 1,3-butadiene ⟶ 4-vinylcyclohexene
    (a) 1. acrolein, sodium methoxide, 2. methyl Grignard
    (b) 1. ethylene, 2. palladium, nitrogen
    (c) heat
    (d) acetylene, heat
    (e) 1. cyclohexene, 2. potassium permanganate, 3. hydrogen, platinum

11. benzaldehyde ⟶ 2-phenylindole
    (a) indole, sodium amide
    (b) 1. aniline, 2. sulfuric acid
    (c) 1. o-toluidine, 2. sulfuric acid
    (d) 1. phenylhydrazine, 2. acid
    (e) 1. o-aminobenzaldehyde, 2. sulfuric acid

12. butyric acid ⟶ n-propylamine
    (a) 1. thionyl chloride, 2. sodium amide
    (b) 1. sodium permanganate, 2. thionyl chloride, 3. ammonia
    (c) 1. thionyl chloride, 2. ammonia, 3. bromine, aqueous sodium hydroxide
    (d) 1. bromine, 2. potassium phthalimide, 3. aqueous acid
    (e) 1. dicyclohexylcarbodiimide, 2. ammonia

13. cis-3,4-dimethylcyclobutene ⟶ cis,trans-1,4-dimethyl-1,3-butadiene
    (a) aluminum chloride
    (b) platinum oxide, nitrogen
    (c) silver oxide
    (d) heat

(e) light

14. acrolein ⟶ 3-methoxypropanal
   (a) 1. methanol, hydrogen chloride, 2. aqueous acid
   (b) sodium methoxide
   (c) 1. hydrogen peroxide, 2. sodium hydroxide, methyl iodide
   (d) 1. hydrogen chloride, 2. sodium methoxide
   (e) methanol, benzoyl peroxide, light

For each of questions 15-25 choose the answer which <u>best</u> completes the indicated statement.

15. The LUMO of 1,3-butadiene has ____ nodes.
   (a) -1
   (b) 0
   (c) 1
   (d) 2
   (e) 3

16. Sucrose contains ____ chiral centers
   (a) 10
   (b) 6
   (c) 4
   (d) 9
   (e) 12

17. Methyl α-D-ribofuranoside and methyl β-D-ribofuranoside are ____.
   (a) carbohydrates
   (b) reducing sugars
   (c) hormones
   (d) anomers
   (e) esters

18. Enol forms of β-diketones are stabilized by ____.
   (a) π-delocalization
   (b) hydrogen bonding
   (c) dimer formation
   (d) charge dispersal
   (e) the β-effect

19. Dinitrofluorobenzene is of utility in the analysis of ____.
   (a) carbohydrates
   (b) disaccharides
   (c) lipids
   (d) peptides
   (e) polynuclear aromatics

20. The LUMO of 1,3-butadiene contains ____ electrons.
    (a) π
    (b) 2
    (c) σ
    (d) bonding
    (e) 0

21. The pinacol rearrangement proceeds via ____ route.
    (a) an electrophilic substitution
    (b) a free radical
    (c) a cycloaddition
    (d) a carbonium ion
    (e) a carbanion

22. Phenol is increased in acidity by ____.
    (a) increasing its concentration.
    (b) reaction with methanol.
    (c) methyl substitution in the meta position.
    (d) dissolution in pyridine.
    (e) nitro substitution in the para position.

23. Upon alkylation of an enamine, ____ is formed.
    (a) a ketone
    (b) an amine
    (c) an iminium ion
    (d) an alcohol
    (e) an imine

24. The cyclooctatetraenyl dianion is ____.
    (a) stable only in methanol solution.
    (b) non-planar.
    (c) non-absorbing in the UV region.
    (d) aromatic.
    (e) paramagnetic.

25. If the nitrogen of pyrrole is protonated, the ring is ____.
    (a) cleaved.
    (b) no longer aromatic.
    (c) susceptible to electrophilic substitution.
    (d) expanded.
    (e) unaffected.

First Examination (E)   Answer Set

1.(c) 2.(c) 3.(b) 4.(e) 5.(c) 6.(e) 7.(a) 8.(a) 9.(c) 10.(c) 11.(d) 12.(c)

13.(d) 14.(a) 15.(d) 16.(d) 17.(d) 18.(b) 19.(d) 20.(e) 21.(d) 22.(e)

23.(c) 24.(d) 25.(b)

Final Examination (A)   Two and One-Half Hours

1. (25 pts, 20 min)
   (a) Draw a molecular orbital energy diagram for the cyclopentadienyl anion showing all π-electrons present.
   (b) Draw the major resonance structures that would be used to indicate charge delocalization in the cyclopentadienyl anion using valence bond pictures.

2. (25 pts, 15 min) Give brief (two sentences or less) explanations for each of the following. Draw structures where necessary.
   (a) An attempt to form t-butyl methyl ether by the reaction of sodium methoxide with t-butyl bromide failed.
   (b) 2,4-dinitrofluorobenzene may be used in peptide degradative analyses but bromobenzene may not.
   (c) Trichloroacetic acid is a stronger acid than acetic acid.
   (d) The two methyl groups in acetone produce only one signal in its nmr spectrum, whereas the following ketal of acetone exhibits a separate signal for each of the methyl groups.

   $H_3C-C(O-)(O-)-CH_3$ (cyclopentane ketal)

3. (25 pts, 20 min) Suppose one is given a sample of an organic compound and is told that it is an aldotetrose. Give the chemical reactions and briefly describe the rationale involved in determining the absolute configuration of the molecule. Assume that there is available an authentic sample of D-glyceraldehyde.

4. (25 pts, 15 min) The Hell-Volhard-Zelinsky reaction involves α-bromination of a carboxylic acid. Give a detailed mechanism for the overall conversion with the reagents shown.

   $CH_3COOH \xrightarrow{P, Br_2} BrCH_2COOH$

5. (25 pts, 20 min) Show all steps necessary for the synthesis of alanylglycylvaline from the free amino acids.

6. (25 pts, 20 min) Provide the proper reagents for the conversions A-E.

7. (25 pts, 20 min) Devise syntheses for any three of the following compounds starting with diethyl malonate or ethyl acetoacetate and any other needed reagents.
   (a) 2-pentanol
   (b) cyclopentane carboxylic acid
   (c) 2-methyl-4-phenylbutane
   (d) 2,4-pentanedione

8. (25 pts, 20 min) Give the correct structures for compounds A-I.

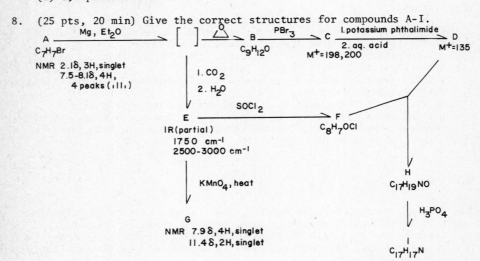

Final Examination (A)   Answer Set

1. (a)

E ↑ ⥮ ⥮ ⥮   — — π*
               π
               π

Six electrons are placed in the three lowest energy π-molecular orbitals. Two anti-bonding molecular orbitals remain unfilled.

(b)

[cyclopentadienyl anion resonance structures showing negative charge delocalized on each of the five ring carbons]

Using simple valence-bond pictures we may see that the negative charge of the species is evenly distributed over the five carbons of the ring system. As you should remember from the concepts of charged-particle energetics as presented in introductory chemistry, spreading a charge over a larger region of space causes its energy to be lowered.

2. (a) In the tertiary halide system, treatment with a strong base leads almost entirely to alkene. That is, elimination predominates.

(b) This method of peptide analysis uses a nucleophilic aromatic substitution reaction. Such a reaction is facilitated by the presence of two strongly electron-withdrawing groups of the 2,4-dinitrofluorobenzene which are absent in the simple bromobenzene system.

(c) The combined electron-withdrawing effect of three chlorines facilitates proton removal in aqueous solution by decreasing the negative charge at the site from which the proton departs.

(d) The two methyl groups in question are held rigidly in non-equivalent positions because of the rigidity of the ring system and the tetrahedral nature of the carbon to which they are bound.

3. There are four possibilities to be considered:

```
   CHO           CHO           CHO           CHO
 ⊢ OH         HO ⊣          ⊢ OH         HO ⊣
 ⊢ OH         HO ⊣         HO ⊣           ⊢ OH
  CH₂OH        CH₂OH         CH₂OH         CH₂OH
    A            B             C             D
```

Species A and B might be eliminated or confirmed by a nitric acid oxidation to a diacid. The diacid would be optically active if structure C or D were present, and optically inactive if the structures were A or B. Further distinction between A and B (or between C and D) may be done by degradation to glyceraldehyde (for example, 1. NH_2OH; 2. Ac_2O; 3. NH_3; 4. Ag_2O) and comparison of the product to the authentic sample available.

4.

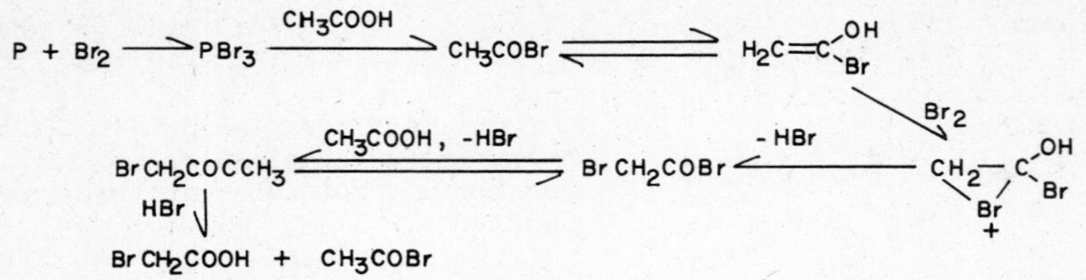

$$P + Br_2 \longrightarrow PBr_3 \xrightarrow{CH_3COOH} CH_3COBr \rightleftharpoons H_2C=C\begin{smallmatrix}OH\\Br\end{smallmatrix}$$

$$\xrightarrow{Br_2} CH_2\begin{smallmatrix}\\Br\end{smallmatrix}-C\begin{smallmatrix}OH\\Br\end{smallmatrix} \xrightarrow{-HBr} BrCH_2COBr \xrightleftharpoons[]{CH_3COOH,\ -HBr} BrCH_2COCCH_3$$

$$\downarrow HBr$$

$$BrCH_2COOH + CH_3COBr$$

Several aspects of this sequence are critical. First, the initial reaction is to form PBr_3 which generates the acyl halide. The latter is involved in a keto-enol type equilibrium which generates the enol form in quantity sufficient that bromination may occur. This is postulated to proceed <u>via</u> a bromonium type intermediate, analogous to alkene bromination. Conversion to the final acid is by an equilibrium with a mixed anhydride.

5. The fundamental approach in this problem is similar to one discussed previously. In addition to the method illustrated, coupling could be performed by another activation method, as could protection of the amino terminus. The mixed anhydride method of activation is shown here.

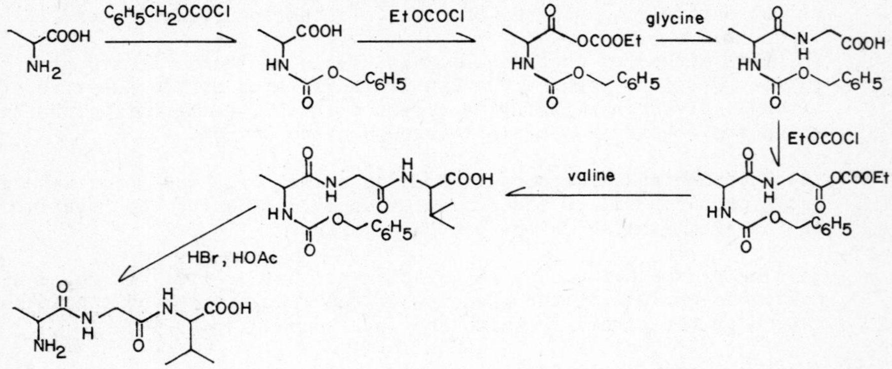

[Scheme: alanine (COOH, NH₂) → treated with $C_6H_5CH_2OCOCl$ to give Cbz-alanine → EtOCOCl forms mixed anhydride (OCOOEt) → glycine couples to give Cbz-Ala-Gly-COOH → EtOCOCl activates to mixed anhydride → valine couples to give Cbz-Ala-Gly-Val-COOH → HBr, HOAc deprotects to give Ala-Gly-Val tripeptide]

6. (A) Br_2, HOAc
The reaction must be performed under acidic conditions. Under basic conditions polyhalogenation and displacement will occur.

(B) Br_2, KOH
Performing the halogenation under basic conditions leads to the haloform reaction with generation of the carboxylic acid.

(C) PBr_3, Br_2
This synthesis should be recognized as the Hell-Volhard-Zelinsky reaction for which the mechanism involves the bromination of the enol form of the acyl bromide.

(D) 1. SOCl$_2$, 2. (CH$_3$)$_2$Cd
This is the preferred route for the direct conversion of a carboxylic acid to a ketone.

(E) 1. SOCl$_2$, 2. NH$_3$, 3. P$_2$O$_5$
It should be noted that a nitrile is really a derivative of a carboxylic acid, that is, a dehydrated amide. The route here involves dehydration using an inorganic anhydride.

7. (a) [structure: OH compound →(LiAlH$_4$) ← ketone →(aq. acid, heat) ← →(1. NaOEt, 2. EtI) ← COOEt compound]

In general, for syntheses of acetoacetic ester or malonic ester types, substituents (alkyl groups) on the position α to the carbonyl group (oxygen-bearing carbon) are introduced in the initial stages. In this system the target molecule bears an ethyl group in this position. Thus ethyl iodide is used initially.

(b) [cyclopentane-COOH →(aq. acid, heat) ← cyclopentane(COOEt)$_2$ →(NaOEt) ← alkylated cyclopentane diester ←(NaOEt) Br-CH$_2$CH$_2$-Br + EtOOC-CH$_2$-COOEt]

This represents a two-step malonic ester synthesis, the second being an intramolecular process.

(c) [H$_5$C$_6$ compound →(H$_2$, Pd) ← alkene (H$_5$C$_6$) →(1. CH$_3$MgI, 2. aq. acid) alcohol (H$_5$C$_6$) →(aq. acid, heat) ketone (H$_5$C$_6$) ← β-ketoester (H$_5$C$_6$, COOEt) ←(1. NaOEt, 2. C$_6$H$_5$CH$_2$Br) acetoacetic ester COOEt]

One should recognize readily the need for acetoacetic ester rather than diethyl malonate due to the lack of an oxygenated function in the product. The rest of the solution should be a familiar story by now. One must add the two alkyl groups, and one should recognize that attempting to put the methyl on by the acetoacetic ester synthesis is foolhardy as the remainder of the route would be excessively long.

(d) [diketone ←((CH$_3$)$_2$Cd) Cl-CO-CH$_2$-CO-? ←(SOCl$_2$) HOOC-CH$_2$-CO-? ←(aq. acid) EtOOC-CH$_2$-CO-?]

No doubt one could devise other routes using the classical approach of the acetoacetic ester synthesis, complete with decarboxylation. However, these would be much longer. In practice, additional steps may add weeks or months to the job rather than the few seconds it takes to write them.

8.

A: 4-bromotoluene (methylbenzene with Br para)
B: 4-methylbenzyl alcohol (CH₂OH para to methyl)
C: 4-methylbenzyl bromide (CH₂Br para to methyl)
D: 4-methylbenzylamine (CH₂NH₂ para to methyl)
E: p-toluic acid (COOH para to methyl)
F: p-toluoyl chloride with HOOC (terephthalic acid mono acid chloride — COCl and HOOC para)
G: terephthalic acid derivative — COOH para-substituted (HOOC and COOH)
H: amide — 4-methylbenzyl group NH-C(=O)-C₆H₄-CH₃ (para-methyl)
I: 1-(4-methylphenyl)-6-methyl-3,4-dihydroisoquinoline

The greatest information for this sequence is given at the beginning with A. The structure is obtained from the formulation and the nmr (a methyl group; the AA'BB' pattern indicating para substitution of unlike groups). The remainder is fairly straightforward, but requires knowledge of the variety of reactions involved. The nmr of G is indicative of a para disubstituted system with identical groups. Be careful of the substitution pattern when piecing together the isoquinoline ring.

Final Examination (B)   Two and One-Half Hours

1. (25 pts, 20 min) Explain each of the following:
    (a) The compound 6,6'-dibromobiphenyl-2,2'-dicarboxylic acid is obtainable in optically active forms in spite of the fact that the molecule contains no chiral atom.
    (b) Cyclopentadiene is "unusually" acidic for a hydrocarbon species.
    (c) Upon treatment with sodium nitrite and sulfuric acid, N,N-dimethylaniline yields a product in which the amino nitrogen is unchanged.

2. (25 pts, 15 min) Write reactions using real compounds (that is, actual structures, not R-, etc.) of five carbons or more illustrating the use of each of the following reagents. Products must be isolable and any pertinent stereochemistry illustrated.
    (a) acetic acid, hydrogen peroxide
    (b) phosphorus tribromide, bromine
    (c) carbon dioxide
    (d) sodium cyanide
    (e) bromine, potassium hydroxide, water

3. (25 pts, 20 min)
    (a) Using a molecular orbital approach, explain the stability of the cycloheptatrienyl cation and the cyclobutadienyl dication.
    (b) In the UV trans-stilbene exhibits a λ_{max} at 295 nm with ε = 27,000 while cis-stilbene exhibits a λ_{max} at 280 nm with ε = 13,000. Rationalize this result.
    (c) On treatment with triethylamine in ether, propionyl chloride generates a solid precipitate. After filtration and addition of ethanol to the filtrate, ethyl propionate is formed. Explain what is happening.

4. (25 pts, 20 min) Devise syntheses for any three of the following compounds starting with the indicated materials and using any other organic and inorganic reagents thought necessary.
    (a) 4-phenylquinoline from phenyl phenyl ketone (benzophenone)
    (b) 1-(α-naphthyl)isoquinoline from α-naphthoyl chloride
    (c) 2-(1-carboxyethyl)phenol from phenol

(d)

   ⌬-OH   from   ⌬-NO₂

5. (25 pts, 15 min) A tripeptide isolated from a biochemical preparation was suspected of being alanylglycylalanine. To prove it, a degradative study was performed. Show all reagents and products that would be used and obtained in such a degradation, assuming the suspected structure was right.

6. (25 pts, 20 min) Complete the following reaction scheme by giving the correct structural formulas for the materials designated A-G.

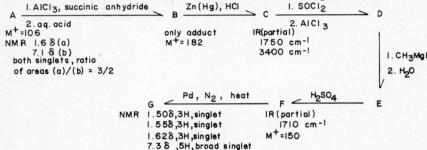

A $\xrightarrow{\text{1. AlCl}_3,\ \text{succinic anhydride}}_{\text{2. aq. acid}}$ B $\xrightarrow{\text{Zn(Hg), HCl}}$ C $\xrightarrow[\text{2. AlCl}_3]{\text{1. SOCl}_2}$ D

$M^+ = 106$
NMR 1.6 δ (a)
   7.1 δ (b)
both singlets, ratio
of areas (a)/(b) = 3/2

only adduct
$M^+ = 182$

IR (partial)
1750 cm⁻¹
3400 cm⁻¹

D $\xrightarrow[\text{2. H}_2\text{O}]{\text{1. CH}_3\text{MgI}}$ E

G $\xleftarrow{\text{Pd, N}_2,\ \text{heat}}$ F $\xleftarrow{\text{H}_2\text{SO}_4}$ E

G NMR 1.50 δ, 3H, singlet
        1.55 δ, 3H, singlet
        1.62 δ, 3H, singlet
        7.3 δ, 5H, broad singlet

F IR (partial)
   1710 cm⁻¹
   $M^+ = 150$

7. (25 pts, 20 min) Decarboxylation of an "ordinary" aliphatic carboxylic acid is often difficult. The following Hunsdiecker reaction is useful for this purpose. Suggest a mechanism for the process.

$$RCH_2CH_2COO^-Ag^+ \xrightarrow{Br_2,\ CCl_4} RCH_2CH_2Br + AgBr + CO_2$$

8. (25 pts, 20 min) Show all reagents (and isolable intermediate products) in the following conversions.

(a) [methylcyclopentanone] → HOOC-CH(CH₃)-CH₂-CH₂-COOH

(b) CH₃-CO-CH₂-CN → HOOC-CH=CH-C₆H₅

(c) C_6H_5CHO → [tetrahydropyran with two C₆H₅ groups and O]

(d) (CH₃)₃C-COOEt → HO-C(CH₃)₂-... -COOEt

Final Examination (B)   Answer Set

1. (a) Owing to the bulky substituents in each ring in the position ortho
    relative to the bond joining the rings, rotation about this bond is
    severely hindered. Thus the molecule may be held in an optically active
    conformation. The mirror images of such a system are not superimposable
    as may be seen in the following diagram.

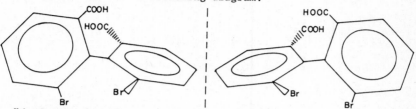

(b) Charge delocalization causes significant stabilization for the cyclopentadienyl anion. An aromatic system is generated, one of a cyclic π
system bearing (4n+2) π electrons. This is illustrated with the following
resonance structures and M.O. diagram:

(c) The aromatic ring in this system is activated for electrophilic substitution and the nitrosonium ion attacks in the para-position.

2. (a) [alkene] + HOAc, H$_2$O$_2$ → [epoxide]

    (b) [isobutyric acid] + PBr$_3$, Br$_2$ → [α-bromo acid]

    (c) [RMgBr] + CO$_2$ → [R-COO$^-$]

    (d) [R-Br] + NaCN → [R-CN]

(e) [structure: isopropyl methyl ketone] →(Br₂, KOH, H₂O) [structure: isopropyl COO⁻]

or

[structure: α-amino isobutyramide] →(Br₂, KOH, H₂O) [structure: isopropyl NH₂]

3. (a) The following simple energy diagrams indicate that each of the ions have completely filled π-bonding molecular orbitals with no extraneous electrons. Moreover, these now fit the Huckel aromaticity rule since they possess (4n+2) π electrons in a cyclic π system.

(b) In the case of <u>cis</u>-stilbene there is a steric interaction between the phenyl rings which prevents the full delocalization of electrons in the π-system by forcing a twisting of the rings from coplanarity. Thus the <u>cis</u>-stilbene absorbs higher energy light (shorter wavelength) than the <u>trans</u>-stilbene for which there is no such steric interaction. Accompanying this shift in energy is a drop in the efficiency of excitation of the electron system as evidenced by the reduced extinction coefficients.

(c) The precipitate is triethylamine hydrochloride which is generated as a by-product in the dehydrohalogenation of the starting acid chloride to give a ketene

$$CH_3CH_2COCl \xrightarrow{Et_3N} CH_3CH=C=O$$

This "monomeric anhydride" of propionic acid is quite reactive and readily forms an ester upon addition of an alcohol.

4. (a) [structure: 4-phenylquinoline] ⇐(aniline, H₃PO₄, phenyl vinyl ketone)

This is a modification of the Skraup synthesis in which the unsaturated carbonyl compound is added directly rather than generated <u>in situ</u>.

180

(b)

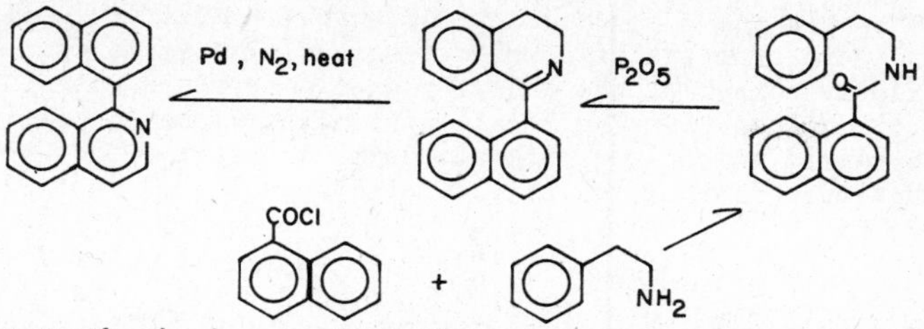

[Scheme showing conversion of naphthyl acid chloride + phenethylamine → amide → (P₂O₅) dihydroisoquinoline → (Pd, N₂, heat) aromatized isoquinoline]

This synthesis is an adaptation of the classic Bischler-Napieralski method for the preparation of an isoquinoline ring system. The acid-catalyzed ring closure is performed here with P_2O_5 and aromatization by dehydrogenation over a standard hydrogenation catalyst, the product hydrogen being swept away with nitrogen.

(c) An overall view of this problem reveals that an <u>ortho-specific</u> substitution on phenol is necessary. Such a reaction is the <u>Reimer-Tiemann</u> process which places an aldehyde function on the ring. This is then functionalized, as shown in the following scheme, by use of the Reformatsky reaction.

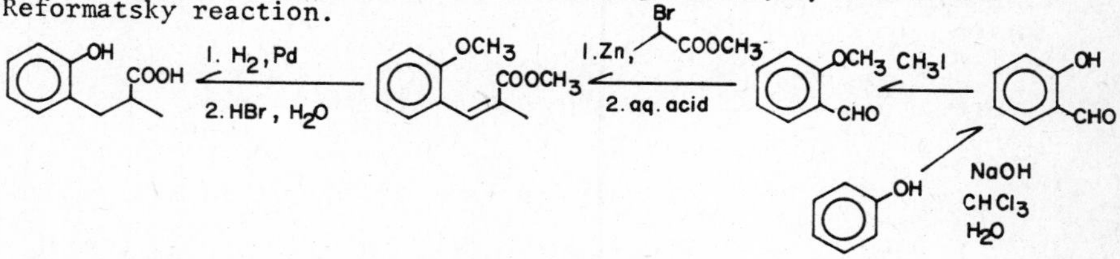

[Scheme: phenol → (NaOH, CHCl₃, H₂O) salicylaldehyde → (CH₃I) o-methoxybenzaldehyde → (1. Zn, BrCH₂COOCH₃; 2. aq. acid) → methyl ester with OCH₃ → (1. H₂, Pd; 2. HBr, H₂O) → final o-hydroxyphenylpropionic acid]

(d)

[Scheme: nitrobenzene → (Sn, HCl) aniline → (1. NaNO₂, H₂SO₄; 2. Cu₂(CN)₂) benzonitrile → (aq. acid) benzoic acid → (LiAlH₄) benzyl alcohol]

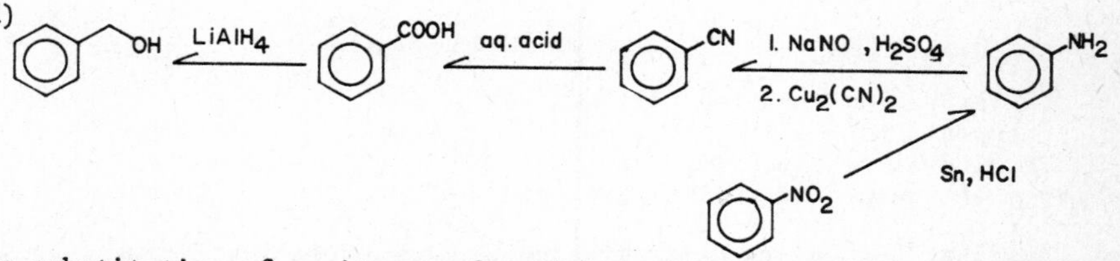

The substitution of a nitrogen function on an aromatic ring is required with the formation of a carbon-carbon bond. Several routes are possible once the diazonium ion is generated. The one shown here requires the fewest steps. A Sandmeyer reaction is performed followed by hydrolysis and reduction.

5. 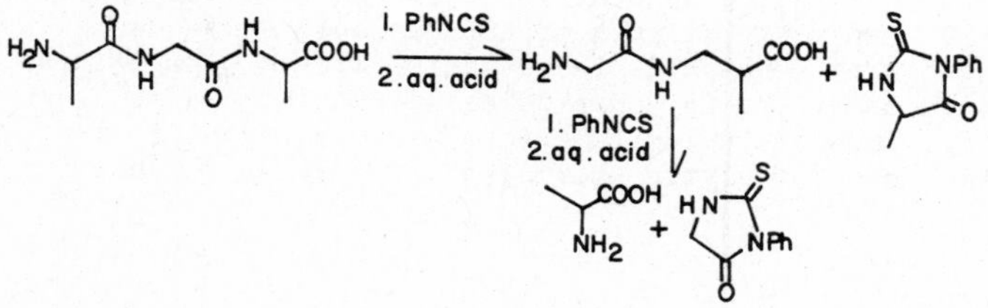 [Scheme: tripeptide H₂N-CH(R)-CO-NH-CH₂-CO-NH-CH(CH₃)-COOH → (1. PhNCS; 2. aq. acid) → dipeptide H₂N-CH₂-CO-NH-CH(CH₃)-COOH + phenylthiohydantoin of first residue; then (1. PhNCS; 2. aq. acid) → alanine (H₂N-CH(CH₃)-COOH) + phenylthiohydantoin of glycine]

The route shown is the Edman degradation which removes successively a single amino acid from the free amino terminus. The amino acid removed is identified by comparing the phenylhydantoin obtained with that prepared from authentic amino acid. The final amino acid may be identified directly by chromatographic comparison with authentic material. Other routes are possible, but this is the simplest definitive one.

6.

A: p-xylene
B: 3-(o-tolyl)-3-oxopropanoic acid (o-methylphenyl ketone with CH₂COOH)
C: 3-(o-tolyl)propanoic acid
D: 5,8-dimethyl-1-tetralone
E: 1-hydroxy-tetralin (1,2,3,4-tetrahydronaphthalen-1-ol, dimethyl substituted)
F: 1,2-dihydronaphthalene (dimethyl)
G: dimethylnaphthalene

The clues for the point of attack in this problem are in the spectrometric data for $\underline{A}$ and the extra note concerning $\underline{B}$. From the nmr and mass spectral data for $\underline{A}$ one concludes that $\underline{A}$ is a xylene, although which might still be in doubt. This doubt is removed by the generation of only one <u>mono</u> substitution product.

7.

$$RCH_2CH_2COOAg \xrightarrow{Br_2,\ CCl_4} RCH_2CH_2COOBr + AgBr$$

$$RCH_2CH_2\cdot + CO_2 \longleftarrow RCH_2CH_2C(O\cdot)=O + Br\cdot$$

$$Br_2 \searrow \quad \searrow RCH_2CH_2COOBr$$

$$RCH_2CH_2Br + Br\cdot \qquad RCH_2CH_2Br + RCH_2CH_2C(=O)O\cdot$$

This problem was probably found to be difficult unless the student had happened to meet it previously. The formation of silver bromide should be a clue as to the nature of the first step of the mechanism, and therefore to the nature of the original brominated organic. Since Br^- is produced from Br_2, an oxidation-reduction must occur that requires combination of the carboxylate anion with an electron-deficient bromine species (in effect, Br^+). This would generate the brominated intermediate RCH_2CH_2COOBr. Alternate ways of looking at this initial process lead to the same structure for the original brominated organic. This must then undergo some type of decarboxylation process. One might try to write a completely ionic mechanism to account for this decarboxylation, but it should be recalled that only if loss of CO_2 leads to a resonance stabilized carbanion is such an ionic mechanism likely. Furthermore, a completely ionic mechanism in a non-polar solvent such as CCl_4 is somewhat unlikely. Thus one needs to invoke a free-radical chain mechanism, initiated by homolytic thermolysis of the oxygen-bromine bond.

8. (a) [reaction scheme: dimethyl-substituted succinic acid with two COOH groups → KMnO4 → methylcyclopentene → H2SO4 → methylcyclopentanol → NaBH4 or LiAlH4 ← methylcyclopentanone]

One should realize that the same number of carbon atoms is present in the starting material and the target molecule. The fundamental reaction type is one we have met before.

(b) HOOC—CH=CH—C6H5 → KOH, I2, H2O → PhCH=CH—CO—CH3 ← aq. acid, heat ← PhCH2—CH(CN)—CO—CH3

with branch: CH3—CO—CH2—CN + PhCHO → NaOEt

Again, the sequence is almost identical to one we have met previously.

(c) H5C6—(tetrahydropyranone) ← aq. acid ← H5C6—CH(—)—CH2—CH(OH)—CH2—COOH ← NaBH4 ← H5C6—CH(—)—CH2—CO—CH2—COOH

from: 1. EtOOCCH2COOEt, NaOEt; 2. aq. acid, heat

PhCHO → acetone, base → H5C6—CH=CH—CO—CH3

This is a difficult problem requiring the use of a number of reaction types. It is essential to work backwards from the product in problems like this. The desired product is a lactone which is made by acid catalyzed dehydration of the appropriate hydroxy-carboxylic acid. Examination of the carbon skeleton of the hydroxy acid reveals that it is accessible from benzaldehyde by use of a condensation reaction with acetone, a Michael addition with malonic ester, and finally NaBH4 reduction.

(d) (CH3)2C(OH)—CH2—COOEt ← Zn, BrCH2COOEt ← (CH3)2C—CHO ← LiAlH(O-t-Bu)3 ← (CH3)2C—COCl ← SOCl2 ← (CH3)2C—COOH

then → aq. acid → (CH3)2C—COOEt

The route illustrated makes use of the Reformatsky reaction for the formation of the new carbon-carbon bond present in the target molecule. Beyond this, the major task is conversion of the starting material into the corresponding aldehyde. The student might be tempted to write as the solution a crossed Claisen condensation of the given starting ester with ethyl acetate. However, this would probably undergo dehydration, and specific reintroduction of the hydroxyl would be difficult.

183

Final Examination (C)   Two and One-Half Hours

1. (25 pts, 20 min) A molecular orbital diagram shows the planar cyclooctatetraene dianion to be a stable aromatic species, but the parent (planar) molecule to be a diradical. Draw the appropriate molecular orbital diagrams. Cyclooctatetraene is a stable molecule (that is, it can be put in a bottle and stored) in spite of the molecular orbital diagram. Explain this phenomenon.

2. (25 pts, 20 min) Supply the reagents necessary for the conversions A-E. Please note that more than one step may be required for each conversion.

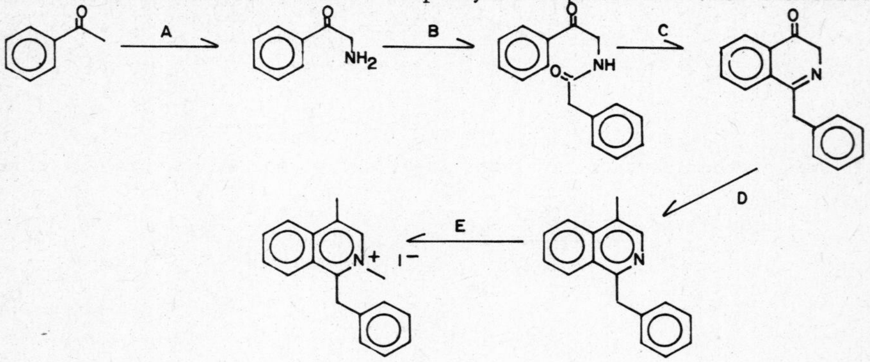

3. (25 pts, 15 min) Predict the major organic product and draw its structure for each of the following reactions.
   (a) furfural + Ac$_2$O, NaOAc
   (b) H$_3$CO-C$_6$H$_4$-CH(CH$_3$)$_2$ O$_2$, aq. acid
   (c) 4-chloro-nitrobenzene + KOCH$_3$
   (d) H$_3$CO-C$_6$H$_4$-NH$_2$ 1. NaNO$_2$, H$_2$SO$_4$ 2. H$_3$PO$_2$, H$_2$O

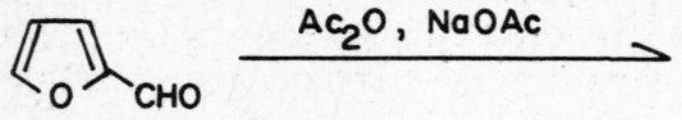

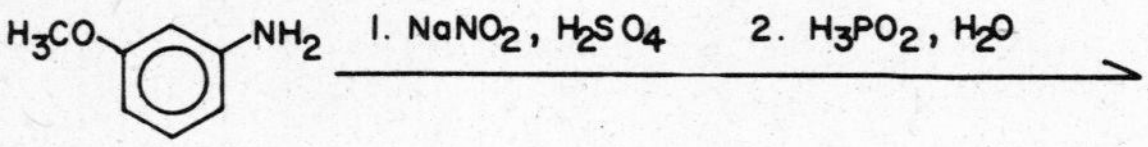

4. (25 pts, 20 min) Suggest mechanisms for the following transformations.
   (a) PhC(O)NH$_2$ NaOH, H$_2$O, Br$_2$ → PhNH$_2$

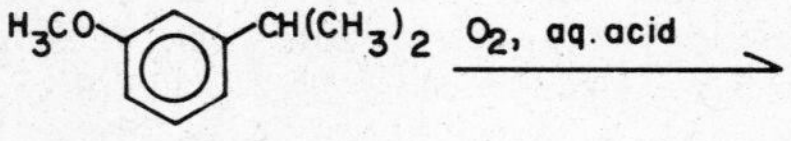

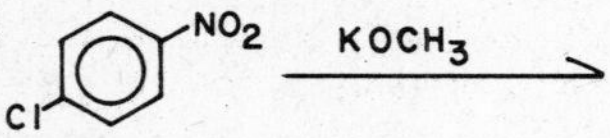

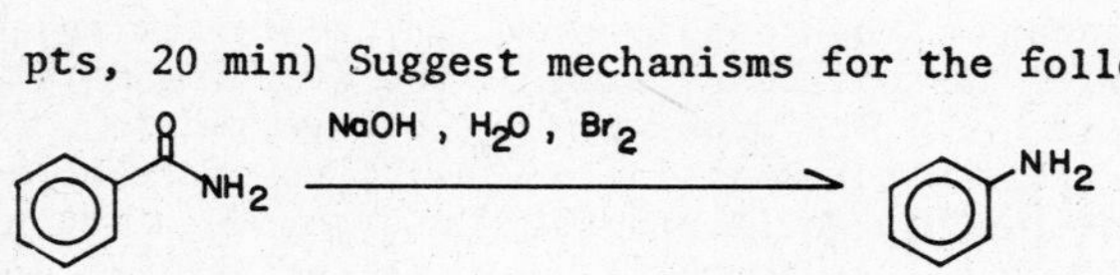

(b)

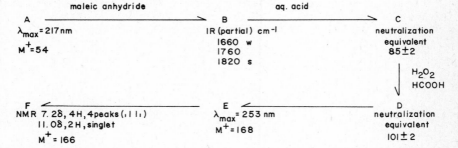

5. (25 pts, 15 min) Outline all steps in <u>two independent</u> syntheses of racemic leucine.

   (CH₃)₂CHCH₂CH(NH₂)COOH

6. (25 pts, 20 min) Devise syntheses for <u>any four</u> of the following compounds from the indicated starting materials using any additional aliphatic and inorganic reagents needed.
   (a) 4-iodo-3-nitrotoluene from toluene
   (b) 4-bromoethylbenzene from toluene
   (c) 3-bromo-4-methylbenzoic acid from toluene
   (d) 2,4-diaminophenol from benzene
   (e) p-methylbenzylamine from toluene

7. (25 pts, 20 min) Give structures for compounds A-F. Be careful of stereochemistry; if a racemic mixture is present, so indicate.

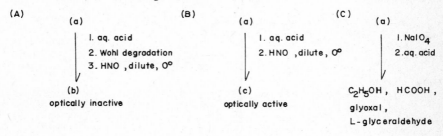

8. (25 pts, 20 min) Given the following pieces of data (A-C) concerning reactions of compound (a), deduce as much structural information as possible about (a). Draw as complete a structure as possible showing all stereochemical aspects and give a brief explanation of the use of each of items (A-C) in writing this structure.

(A) (a)
   1. aq. acid
   2. Wohl degradation
   3. HNO , dilute, 0°
   ↓
   (b) optically inactive

(B) (a)
   1. aq. acid
   2. HNO , dilute, 0°
   ↓
   (c) optically active

(C) (a)
   1. NaIO₄
   2. aq. acid
   ↓
   C₂H₅OH, HCOOH, glyoxal, L-glyceraldehyde

185

Final Examination (C)   Answer Set

1.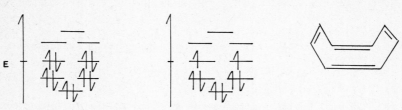

for dianion   for "planar" cyclooctatetraene

The cyclooctatetraene molecule is able to distort from the planar form that is required for the above molecular orbital diagram to have any meaning. In this distorted form (shown below) the π orbitals associated with the individual alkene bonds do not interact appreciably, and the molecule behaves like an ordinary alkene. In this form the molecule is not a diradical.

2. (A) 1. Br_2, HOAc   2. Potassium phthalimide   3. H_3O^+

The potassium phthalimide route is highly favored over reactions with ammonia. Too many things can go awry in the ammonia system, such as formation of higher amines and reaction with the carbonyl group.

(B) ⌬-COCl

(C) P_2O_5

One should recognize this as the Bischler-Napieralski reaction, a type of electrophilic aromatic substitution. Given the reagents, one should be able to postulate a mechanism.

(D) 1. CH_3MgI 2. H_3O^+

This is alkylation followed by dehydration yielding an aromatic system.

(E) CH_3I

As a quaternary ammonium compound is desired here, nothing can go wrong in using the simple alkylation method.

3. (a) ⟨furan⟩-CH=CH-COOH

This is an example of the Perkin condensation, one of many modifications of the aldol condensation. It involves an aldehyde devoid of α-hydrogens, an anhydride (which is the enolate anion source), and a base which is the carboxylate anion derived from the same acid as the anhydride.

(b) 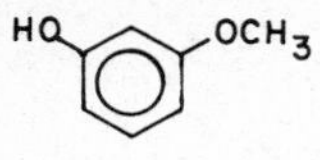 HO-⟨benzene⟩-OCH₃ (meta)

(c) 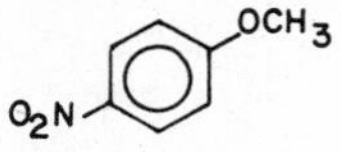 O₂N-⟨benzene⟩-OCH₃ (para)

The presence of the strongly electron-withdrawing nitro group in a position **para** to the halogen causes the halogen site to become susceptible to nucleophilic substitution, here by alkoxide ion.

(d) 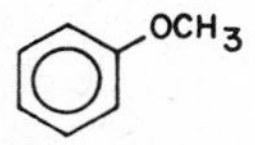 ⟨benzene⟩-OCH₃

One should recognize the reaction conditions for diazotization followed by reduction.

4. (a) This reaction is generally known as the Hofmann rearrangement.

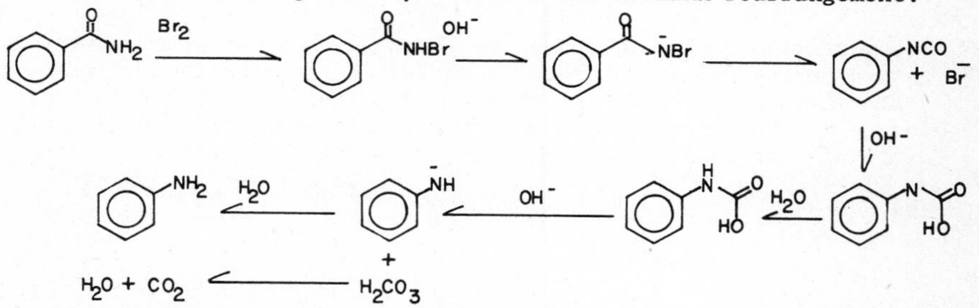

PhC(O)NH₂ →(Br₂) PhC(O)NHBr →(OH⁻) PhC(O)NBr⁻ → PhNCO + Br⁻

↓ OH⁻

PhNH₂ ←(H₂O) PhNH⁻ ←(OH⁻) PhN(H)C(O)OH ←(H₂O) PhN=C=O (with HO)

H₂O + CO₂ ← H₂CO₃

(b) The reaction conditions are those of diazotization, but aliphatic diazonium ions are unstable. The products here indicate that a skeletal rearrangement typical of a carbonium ion type mechanism has taken place. This is readily rationalized by invoking initial loss of N₂ from the diazonium ion followed by rearrangement of a primary to a secondary carbonium ion.

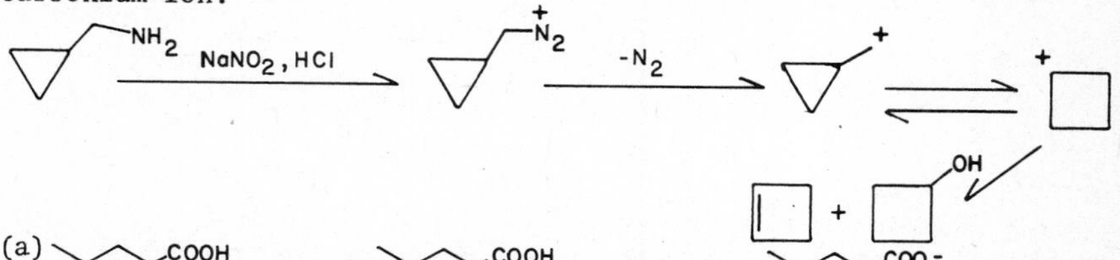

cyclopropyl-CH₂-NH₂ →(NaNO₂, HCl) cyclopropyl-CH₂-N₂⁺ →(-N₂) cyclopropyl-CH₂⁺ ⇌ cyclobutyl⁺

cyclobutyl⁺ + cyclobutanol

5. (a) 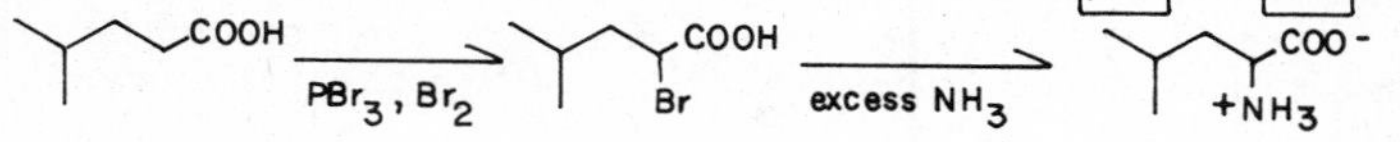 (CH₃)₂CHCH₂CH(CH₃)COOH →(PBr₃, Br₂) (CH₃)₂CHCH₂C(Br)(CH₃)COOH →(excess NH₃) (CH₃)₂CHCH₂C(NH₃⁺)(CH₃)COO⁻

(b) (CH₃)₂CHCH(CHO) —NH₃, HCN→ (CH₃)₂CHCH(CN)(NH₂) —H₃O⁺→ (CH₃)₂CHCH(COOH)(⁺NH₃)

Other methods are, of course, available, such as the malonic ester route (with several variations), and the azlactone synthesis. Those shown, amination of an α-halo acid and the Strecker synthesis, are the simplest to write.

6. (a) [3-methyl-4-nitroiodobenzene] ←1. NaNO₂, H₂SO₄; 2. KI— [3-methyl-4-nitroaniline] ←aq. acid— [3-methyl-4-nitroacetanilide] —HNO₃, H₂SO₄→ [acetanilide-NO₂] /Ac₂O

[benzene] —H₂SO₄, HNO₃→ [nitrobenzene] —Sn, HCl→ [aniline]

Incorporation of the iodide by replacement of a group already present should be the last step. Working backwards from the product makes the route obvious. Introduction of the nitro function or introduction of iodide (or replacable function) for <u>hydrogen</u> at the end would be difficult, since the reactions are not specific for that position.

(b) [p-Br-ethylbenzene] ←Br₂, Fe— [ethylbenzene] ←H₂, Pd/C— [styrene] ←aq. acid— [PhCH(OH)CH₃]

[toluene] —1. Br₂, hν; 2. H₂O→ [PhCHO] /CH₃MgI

The student needs to recognize that an ethyl group can be created from the methyl group.

(c) [3-Br-4-COOH benzene] ←Br₂, Fe— [p-COOH benzene] ←aq. acid— [p-CN benzene] ←1. NaNO₂, H₂SO₄; 2. Cu₂(CN)₂— [aniline]

[benzene] —H₂SO₄, HNO₃→ [nitrobenzene] —Sn, HCl→

This is simpler than the two previous parts. One should recognize the wisdom of introducing the bromine last (rather than the carboxyl). The route shown for carboxyl introduction is better than alkylation-oxidation which can't be controlled.

188

(d) [reaction scheme: H2N-C6H3(OH)-NH2 →(Sn, HCl) O2N-C6H3(OH)-NO2 →(O2, aq. acid) O2N-C6H3(NO2)-CH(CH3)2 ←(H2SO4, HNO3) C6H5-CH(CH3)2 ←(Br, AlCl3) C6H6]

Introduction of the phenolic function in the presence of amino groups would be difficult; rather, it should be performed while they are still nitro groups. The cumene hydroperoxide route is shown. It is shorter than the diazotization routes that might be considered.

(e) [reaction scheme: p-CH3-C6H4-NH2 →(LiAlH4) p-CH3-C6H4-CN →(1. H2SO4, NaNO2; 2. Cu2(CN)2) C6H5-NH2 →(Sn, HCl) C6H5-NO2 ←(H2SO4, HNO3) C6H6]

7. [structures A: 1,3-butadiene; B: cyclohexene-dicarboxylic anhydride; C: cyclohexene-dicarboxylic acid; D: cyclohexane-1,2-diol-dicarboxylic acid (HO, HO, COOH, COOH); E: benzene-dicarboxylic acid (COOH, COOH); F: phthalic acid-like (COOH, COOH)]

One should approach this problem from the point where the greatest degree of structural information is given. In this problem, that point is the spectral data for A and B. The UV of A, along with the molecular ion, indicates a four-carbon diene. In its reaction with maleic anhydride, the ir indicates the residual presence of an olefinic linkage, and also an anhydride. This combination of data leads to the conclusion that a Diels-Alder reaction has occurred, and A must be 1,3-butadiene. One should note that a racemic mixture is formed with compound D.

8. From the data (C) one learns that the compound (a) is a derivative of an L-aldohexose, specifically the ethyl glycoside in a pyranose ring. From the data (B) one learns that the derived α,ω-dicarboxylic acid is optically inactive. This knowledge reduces the number of possible parent hydrocarbons to two, L-allose and L-galactose. From the data (A) the second of these is eliminated. Thus the structure is:

[structure: pyranose ring with HO, OH, OH, HO, OH HO, OCH2CH3]

Final Examination (D)   Two and One-Half Hours

1. (25 pts, 20 min) Give the major organic product(s) (if any at all) for each of the following reactions. Be careful of stereochemistry.

   (a) [cyclohexadiene] —heat→

   (b) [cyclohexadiene] + [cyclopentadienone] —heat→

   (c) [diene] —hν→

   Draw the molecular orbitals and show an energy diagram for the ground state of the allyl cation.

2. (25 pts, 15 min) Predict the correct product(s) in each of the following reactions. Be sure to show the proper stereochemistry clearly.

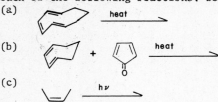

   (4) ←—m-chloroperbenzoic acid / CH_2Cl_2, heat—

   (3) ←—TsOH, $HOCH_2CH_2OH$—

   (7) ←—1. C_6H_5MgBr 2. H_2O—

   (1) —H_2NOH, acid catalyst→

   (2) —H_2, Pd→

   (5) —Ph_3P, CCl_4→

   (6) ←—$KMnO_4$, KOH, H_2O—

3. (25 pts, 15 min) Predict the major organic product in each of the following reactions.

   (a) [N-OH compound] —P_2O_5→

   (b) [cyclohexyl amine] —1. H_2O, 2. heat→

   (c) [hydrazone] —HCl, ethanol→

   (d) [malonate diester] —NaOH, I_2→

   (e) [N-methyl cyclohexyl amine] —1. CH_3I 2. Ag_2O, heat→

4. (25 pts, 20 min) Acetals undergo hydrolysis when treated with aqueous mineral acids, as exemplified in the following reaction. Give the

mechanism of the process in detail.

5. (25 pts, 20 min) Write equations to show how D-glucose could be converted into each of the following:
   (a) methyl 2,3,4,6-tetra-O-methyl-β-D-glucopyranoside
   (b)   (c)

   (Fischer projections shown:
   (b) CHO / HO—H / H—OH / H—OH / CH₂OH
   (c) CHO / HO—H / H—OH / HO—H / H—OH / H—OH / CH₂OH)

6. (25 pts, 20 min) Give the correct structure for each of the compounds indicated A-H.

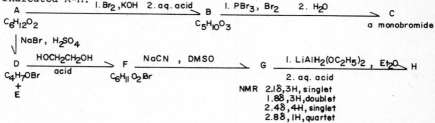

7. (25 pts, 20 min) Give the correct structure for each of the compounds indicated A-F. Be sure to illustrate the proper stereochemistry.

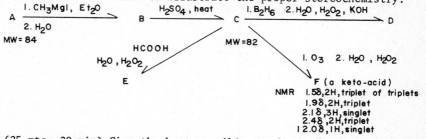

8. (25 pts, 20 min) Give the best possible synthesis for each of the compounds shown below using the indicated starting material and any other reagents thought necessary.
   (a) 1-deutero-2-phenylethane from 2-phenylethanol
   (b) pentanal from 1-bromobutane
   (c) propyl p-tolyl ether from 1-propanol
   (d) ethyl 2-ethoxybutyrate from butanoic acid

Final Examination (D)   Answer Set

1. (a) 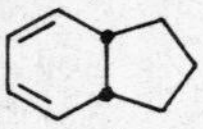 [bicyclic diene structure]

    This is an electrocyclic ring closure of a 6π-electron system proceeding thermally in a disrotatory manner.

    (b) 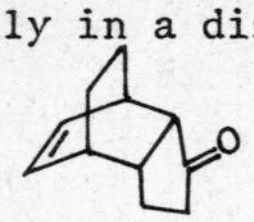 [tricyclic ketone structure]

    This is simply a Diels-Alder reaction, thermally allowed.

    (c) 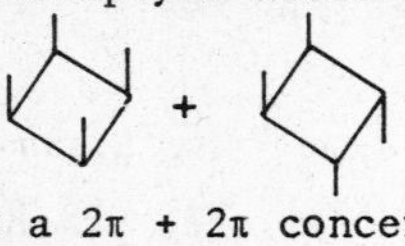 [cyclobutene + cyclobutane structures]

    This is a 2π + 2π concerted cycloaddition which is allowed photochemically

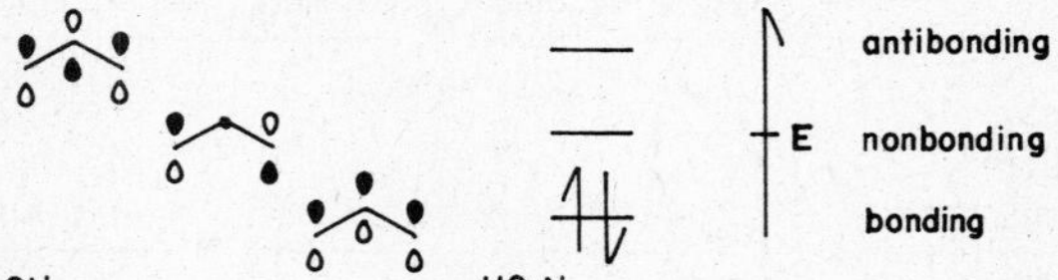

    [orbital diagrams showing bonding, nonbonding, antibonding energy levels]

2. (1) 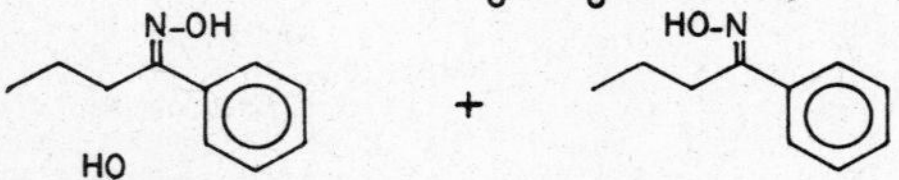 [two oxime structures: N-OH and HO-N isomers] +

    (2) 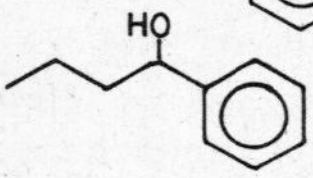 [benzylic alcohol structure] (racemic)

    (3) 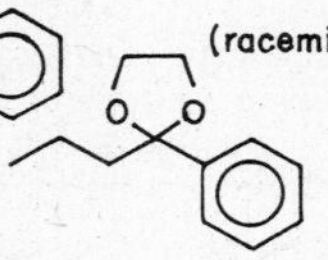 [cyclic acetal (dioxolane) with phenyl and propyl groups]

    (4) 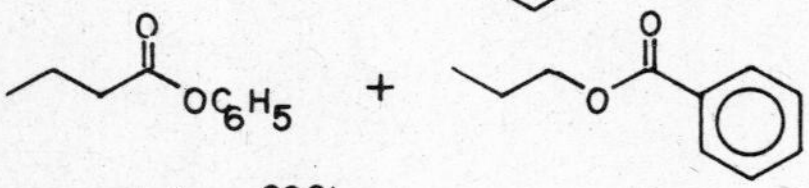 [phenyl butanoate structure] + [propyl benzoate structure]

    (5) 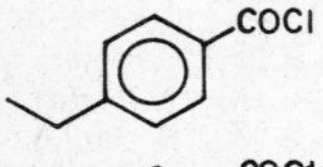 [4-ethylbenzoyl chloride]

    (6) 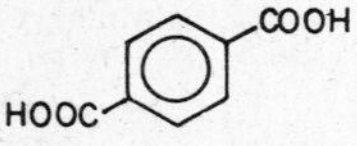 [terephthalic acid, HOOC-C₆H₄-COOH]

    (7) 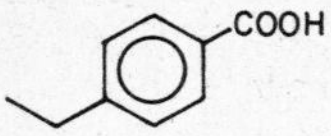 [4-ethylbenzoic acid] (regeneration of starting material)

3. (a) 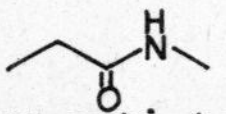 [N-methyl propanamide structure]

    The group <u>anti</u> to the hydroxyl migrates.

192

(b) 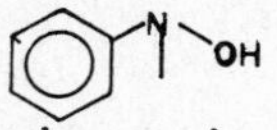 [structure: phenyl-N=CH-OH type]

A β-hydrogen is required for the alkyl function to be eliminated.

(c) 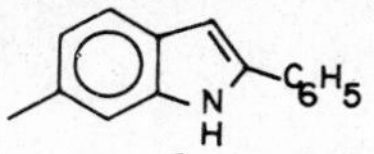 [structure: methylindole with C_6H_5 substituent]

(d) 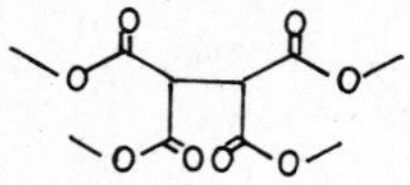 [structure: cyclobutane tetraester]

(e) 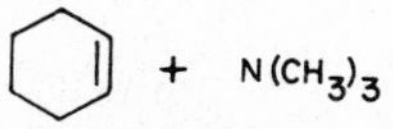 [cyclohexene] + $N(CH_3)_3$

Again a β-hydrogen is required on the alkyl group for elimination.

4. 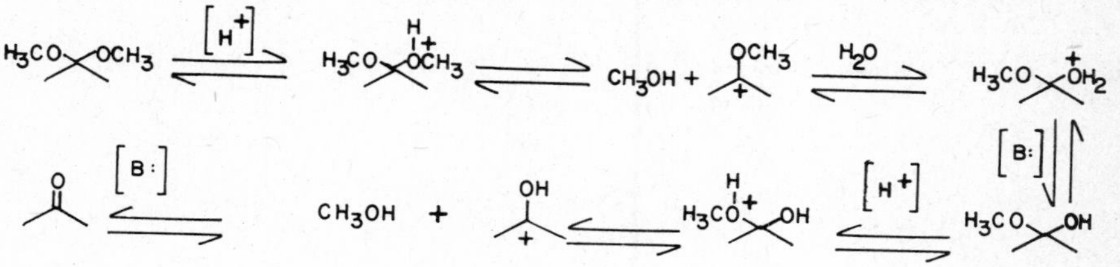 $H_3CO\text{-}C\text{-}OCH_3 \xrightleftharpoons[]{[H^+]} H_3CO\text{-}C\text{-}\overset{+}{O}\overset{H}{CH_3} \xrightleftharpoons{} CH_3OH + \overset{+}{C}\text{-}OCH_3 \xrightleftharpoons{H_2O} H_3CO\text{-}C\text{-}\overset{+}{O}H_2$

$\xrightleftharpoons[]{[B:]} \overset{O}{\underset{}{C}} \xrightleftharpoons{} CH_3OH + \overset{+}{C}\text{-}OH \xrightleftharpoons{} H_3C\overset{+}{O}\text{-}C\text{-}OH \xrightleftharpoons[]{[H^+]} \xrightleftharpoons[]{[B:]} H_3CO\text{-}C\text{-}OH$

The proton source here may be either sulfuric acid itself or the hydronium ion. The base picking up protons may be either bisulfate anion or water.

5. (a)

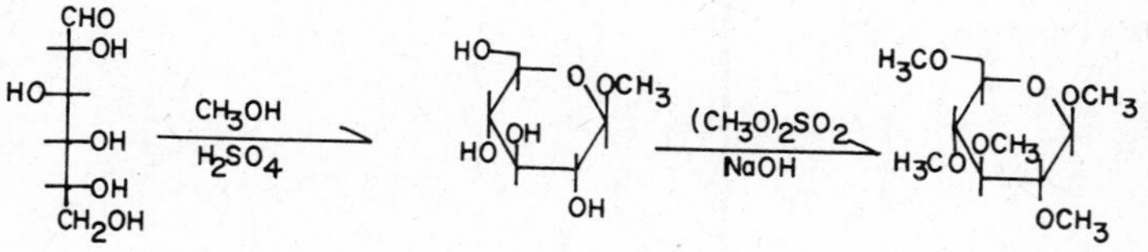

```
  CHO                                              H₃CO
  |-OH                          HO-┐                  ┐-O   OCH₃
HO-|        CH₃OH                  |-O   OCH₃          |
  |-OH      ────→                  |         (CH₃O)₂SO₂  |
  |-OH      H₂SO₄              HO-|                ────→  H₃CO-|    OCH₃
  CH₂OH                            |-OH              NaOH        |
                                                                 OCH₃
```

(b)
```
   CHO                   COOH                  COO⁻ 1/2 Ca⁺⁺
   |-OH                  |-OH                  |=O                     CHO
 HO-|     Br₂, H₂O     HO-|    1. CaCO₃      HO-|    aq. acid        HO-|
   |-OH    ────→         |-OH   ─────→          |-OH   ────→            |-OH
   |-OH                  |-OH   2. Fe⁺⁺⁺,H₂O₂   |-OH    heat            |-OH
   CH₂OH                 CH₂OH                  CH₂OH                   CH₂OH
```

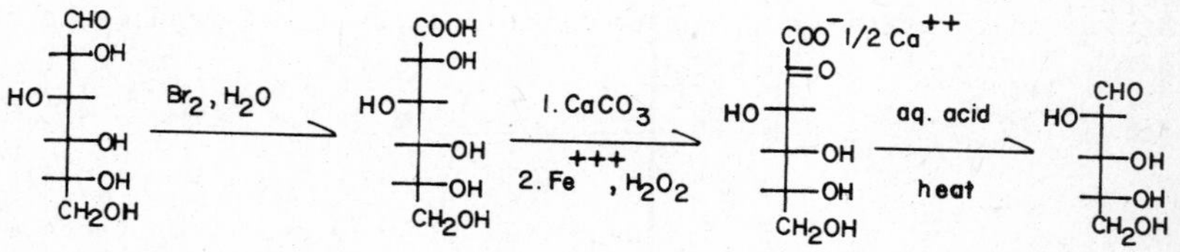

Shown here is the Ruff degradative scheme. The Wohl degradation may also be used.

(c)

<!-- Kiliani-Fischer synthesis scheme -->

```
      CHO                CN            CN              COOH          COOH                         CHO
    ⊢-OH              ⊢-OH          HO-⊣            ⊢-OH          HO-⊣                         HO-⊣
HO-⊣    NaCN      HO-⊣       +  HO-⊣    aq.acid  HO-⊣       +  HO-⊣       1. Ba(OH)₂    HO-⊣
    ⊢-OH  H₂O        ⊢-OH          ⊢-OH   ────>      ⊢-OH          ⊢-OH    ────────>        ⊢-OH
    ⊢-OH              ⊢-OH          ⊢-OH              ⊢-OH          ⊢-OH    2. Na(Hg)          ⊢-OH
     CH₂OH            CH₂OH         CH₂OH             CH₂OH         CH₂OH     aq. acid         CH₂OH
                                                          separate
```

6. A: CH₃COCH(OEt)CH₃ type (ethoxy methyl ketone)
    B: ethoxy-substituted carboxylic acid
    C: α-bromo ethoxy carboxylic acid
    D: α-bromo methyl ketone
    E: ethyl bromide (CH₃CH₂Br)
    F: bromo dioxolane (cyclic acetal with Br)
    G: nitrile dioxolane
    H: 2-methyl-1,3-dicarbonyl (methylmalonaldehyde type)

7. A: cyclopentanone
    B: 1-methylcyclopentanol
    C: methylenecyclopentane (or methylcyclopentene)
    D: methylcyclopentanol
    E: 1-methyl-epoxide of cyclopentane (1-methyl-1,2-epoxycyclopentane)
    F: 5-oxohexanoic acid (keto-acid, HOOC—CH₂CH₂CH₂—CO—CH₃)

The critical structural data here is given with F. Working back from this becomes relatively simple.

8. (a) PhCH₂CH₂D ⟵ 1. Mg, Et₂O ; 2. D₂O ⟵ PhCH₂CH₂Br

(b) OHC—CH₂CH₂CH₂CH₃ ⟵ DMSO/DCC ⟵ HOCH₂—CH₂CH₂CH₂CH₃ ⟵ H₂CO ⟵ BrMg—CH₂CH₂CH₂CH₃ ⟵ Mg, Et₂O ⟵ Br—CH₂CH₂CH₂CH₃

An alternate route is to carbonate the Grignard and reduce the resultant acid.

(c) CH₃CH₂CH₂—O—C₆H₄—CH₃ ⟵ p-CH₃C₆H₄ONa ⟵ CH₃CH₂CH₂—OTs ⟵ TsCl ⟵ CH₃CH₂CH₂OH

Reaction of the alcohol with p-toluenesulfonyl chloride generates a good leaving group for the Williamson route.

(d) CH₃CH(OEt)—COOEt ⟵ NaOEt ⟵ CH₃CHBr—COBr ⟵ Br₂, PBr₃ ⟵ CH₃CH₂COOH

Final Examination (E)	Two and One-Half Hours

Choose the answer which best completes the statement.

1. Ketals are formed from a ketone and two molecules of ____.
   (a) alcohol
   (b) amine
   (c) acid halide
   (d) water
   (e) hydrogen chloride

2. A reducing sugar is one that reacts with ____.
   (a) Fehling's solution
   (b) iodoform
   (c) hydrogen peroxide
   (d) acid halides
   (e) ammonia

3. A nitro group is ____ electrophilic aromatic substitution.
   (a) displaced in
   (b) reduced in
   (c) activating for
   (d) para directing for
   (e) deactivating for

4. The cyclopentadienyl anion has ____ filled π-bonding molecular orbitals.
   (a) 3
   (b) 2
   (c) 4
   (d) 3 degenerate
   (e) 1

5. Amino terminus analysis of a peptide may be done by reaction with ____.
   (a) hydrogen fluoride
   (b) benzophenone
   (c) Fehling's solution
   (d) nitrobenzene
   (e) 2,4-dinitrofluorobenzene

6. In a typical Perkin condensation of an aldehyde with an anhydride, the base is generally ____.
   (a) lithium amide
   (b) sodium ethoxide
   (c) hydroxide ion
   (d) carbonate anion
   (e) a carboxylate anion

7. Benzenediazonium bisulfate is reduced to benzene by reaction with ____.
   (a) hypophosphorus acid
   (b) sodium borohydride
   (c) hydrogen, platinum catalyst

(d) sodium, liquid ammonia
(e) hydrogen sulfide

8. Phenol is less acidic than ____.
   (a) ethanol
   (b) p-methoxyphenol
   (c) p-nitrophenol
   (d) cyclopentadiene
   (e) ammonia

9. Enamines provide a useful route for ____ of ketones.
   (a) reduction
   (b) mono-α-alkylation
   (c) α-oxidation
   (d) hydration
   (e) deoxygenation

10. The thermal dimerization of ethene to cyclobutane is a ____ reaction.
    (a) sodium ion catalyzed
    (b) symmetry allowed
    (c) concerted
    (d) symmetry forbidden
    (e) Diels-Alder type

11. Sucrose is an example of ____.
    (a) a reducing sugar
    (b) a ketohexose
    (c) an aldohexose
    (d) an anomer of ribose
    (e) a disaccharide

12. Grignard reagents react rapidly with carboxylic acids to generate ____.
    (a) esters
    (b) tertiary alcohols
    (c) ketones
    (d) carboxylate anions
    (e) covalently bonded adducts

13. Acetophenone may be monobrominated using ____.
    (a) bromine and sodium hydroxide
    (b) hydrogen bromide in acetic acid
    (c) N-bromosuccinimide
    (d) bromine and acetic acid
    (e) bromine in carbon tetrachloride

14. Aniline is more basic than ____.
    (a) p-methylaniline
    (b) pyrrole
    (c) N-methylaniline
    (d) diethylamine
    (e) ammonia

15. Benzenediazonium chloride and phenol at pH=11 ____.
    (a) do not react
    (b) generate azobenzene
    (c) react explosively liberating nitrogen
    (d) generate anilinium phenoxide
    (e) result in reduction of the diazonium ion

16. Nitrosation of alkyl amines results in the generation of ____.
    (a) carbocations
    (b) alkyldiazonium species as stable entities
    (c) alkylazo compounds
    (d) alkyl nitrites
    (e) amides with carboxylic acids

17. Both the cyclopentadienyl anion and the cycloheptatrienyl cation are ____.
    (a) electron deficient
    (b) aromatic
    (c) readily generated from the corresponding amines
    (d) examples of systems stabilized by hyperconjugation
    (e) volatile as their salts

18. Decahydronaphthalene exists as ____.
    (a) a vapor at room temperature and atmospheric pressure
    (b) cis- and trans- isomers
    (c) a stabilized form of naphthalene
    (d) a hydrated form of naphthalene
    (e) a pair of enantiomers

19. Pyrrole, unlike pyridine, ____ upon protonation.
    (a) loses its aromatic character
    (b) decomposes
    (c) loses volatility
    (d) is increased in reactivity toward electrophiles
    (e) undergoes ring scission

20. Reduction of the aldehyde of D-erythrose yields ____.
    (a) an anomeric carbohydrate
    (b) an optically inactive material
    (c) a pair of enantiomers
    (d) a mixture of three materials
    (e) a cyclic polyol

21. Addition of glycine to distilled water results in ____.
    (a) a slightly acidic solution
    (b) decomposition
    (c) a strongly basic solution
    (d) hydration of the unsaturated function
    (e) a solution of pH=7

22. Tertiary alcohols are ____.
    (a) oxidized under basic conditions
    (b) readily dehydrated under acidic conditions
    (c) more acidic than secondary alcohols
    (d) readily deoxygenated to the saturated hydrocarbon
    (e) all of the above

23. Phenols exhibit a C-O stretching band in the ir which ____.
    (a) is stronger than that for a C=O linkage
    (b) is generally extremely sharp
    (c) is at shorter wavelengths than the corresponding band in alcohols
    (d) occurs in the 1600 cm^{-1} region
    (e) is often observed split to a doublet

24. Cyclohexene with hydrogen peroxide and formic acid results in ____.
    (a) formation of a pair of enantiomers
    (b) epoxide isolation
    (c) a violent explosion due to the liberation of hydrogen
    (d) no reaction
    (e) formation of cyclohexanone

25. Epoxides may be opened by ____.
    (a) aqueous acid
    (b) aqueous base
    (c) reaction with Grignard reagents
    (d) all of the above
    (e) none of the above

26. Carboxylic acids exhibit a characteristic nmr band which is ____.
    (a) split by coupling with the carboxylate carbon
    (b) found at about 12.0 δ
    (c) found at about 7.5 δ
    (d) generally the strongest band in the spectrum
    (e) generally the broadest band in the spectrum

27. Friedel-Crafts acylation proceeds ____.
    (a) most rapidly with nitrobenzene
    (b) only with aromatic acyl halides
    (c) without catalyst when performed on toluene
    (d) *via* initial formation of the acyl anion
    (e) *via* an acylium ion

28. Ammoniacal silver nitrate converts an aldehyde to ____.
    (a) a carboxylic acid
    (b) a primary alcohol
    (c) an acetal in the presence of an acid
    (d) an alkyl nitrate
    (e) a silver complex which precipitates from solution

29. Sodio diethyl malonate addition to acrolein occurs ____.
    (a) at the β-position
    (b) only in acidic solution
    (c) with the generation of an alkoxide ion
    (d) not at all in the absence of peroxides
    (e) with loss of carbon dioxide

30. Lithium aluminum hydride is useful for the reduction of ____.
    (a) aldehydes
    (b) esters
    (c) carboxylic acids
    (d) nitriles
    (e) all of the above

31. The action of base on an alkylphosphonium salt generates an ylide which can react with ____.
    (a) carboxylate esters giving ketones
    (b) activated aryl halides
    (c) carbonyl compounds
    (d) conjugated dienes
    (e) carboxylate esters giving enol ethers

32. Heating an amide with phosphorus pentoxide results in ____.
    (a) liberation of the free amine
    (b) formation of a phosphoramide
    (c) liberation of the free carboxylic acid
    (d) reduction
    (e) formation of a nitrile

33. The anilinium ion exhibits ____.
    (a) no characteristic ir bands
    (b) increased solubility in hexane compared to aniline
    (c) a purple color due to charge delocalization
    (d) decreased reactivity for electrophilic substitution compared to aniline
    (e) a facile decomposition to ammonia

34. Treatment of chlorobenzene with sodium amide results in ____.
    (a) generation of benzyne
    (b) direct displacement of chloride to yield aniline
    (c) electrophilic aromatic substitution
    (d) formation of an arenium ion
    (e) no reaction

35. Acetylacetone is most readily alkylated ____.
    (a) under acidic conditions
    (b) at the central carbon atom
    (c) in ethanol solution
    (d) with tertiary alkyl halides
    (e) with alkyl fluorides

36. The Diels-Alder reaction of 1,3-butadiene proceeds more smoothly with ___.
    (a) ethylene
    (b) irradiation with ir light
    (c) addition of a catalytic amount of lithium amide
    (d) 1,4-pentadiene
    (e) maleic anhydride

37. The β-chloroalkylsulfides undergo hydrolysis involving ___.
    (a) formation of sulfoxides
    (b) direct attack on sulfur
    (c) anchimeric assistance
    (d) carbonium ion formation
    (e) liberation of free elemental sulfur

38. In the cyclopropenyl anion ___ contain electrons.
    (a) the π-bonding molecular orbitals only
    (b) both molecular orbitals
    (c) all of the π-molecular orbitals
    (d) only sigma orbitals
    (e) sigma antibonding molecular orbitals

39. There are ___ stereoisomers in the aldohexose series.
    (a) 8
    (b) 4
    (c) 12
    (d) 32
    (e) 16

40. In a peptide, the carbon-nitrogen bond linking the individual amino acids ___.
    (a) is extremely weak compared to other amide bonds
    (b) is cleaved by catalytic hydrogenolysis
    (c) is readily oxidized
    (d) bears significant double bond character
    (e) is the strongest carbon-nitrogen linkage known

Final Examination (E)   Answer Set

1.(a) 2.(a) 3.(e) 4.(a) 5.(e) 6.(e) 7.(a) 8.(c) 9.(b) 10.(d) 11.(e) 12.(d)

13.(d) 14.(b) 15.(a) 16.(a) 17.(b) 18.(b) 19.(a) 20.(b) 21.(a) 22.(b) 23.(c)

24.(a) 25.(d) 26.(b) 27.(e) 28.(a) 29.(a) 30.(e) 31.(c) 32.(e) 33.(d) 34.(a)

35.(b) 36.(e) 37.(c) 38.(c) 39.(e) 40.(d)